# EXOTIC FRUITS
## A~Z

# Josephine Bacon
# EXOTIC FRUITS A~Z

Line drawings by Soun Vannithone
Photographs by Julian Seaton

*Salem House Publishers*
**TOPSFIELD, MASSACHUSETTS**

DEDICATION: TO HANNA

# ACKNOWLEDGEMENTS

Thanks are due to local vendors of exotic fruits and vegetables, especially Mr N. Patel of West Green Road, London, N.15, the Charalambides family, fruit importers of Caledonian Road, N.7, and Bob Milne of Seven Sisters Market. Many thanks also to my friend, Audrey Ellison, who helped me correctly identify several fruits and vegetables which are easily confused with each other, and to Jack Kessler, of Angmering, Sussex, who lent me some valuable old books on the subject. The author would also like to thank Dr and Mrs Dajani of Bet Hanina and Jericho, for the visit to their orchard, the New Zealand Apple and Pear Council, Dr Randy Keim of the South Coast Field Station, Santa Ana, California, U.S.A., and Avraham Shinar, Director of the Technical Assistance and Foreign Relations Department of the Ministry of Agriculture, Tel Aviv, Israel.

This edition first published in Great Britain by
Xanadu Publications Ltd 1988

First published in the United States by
Salem House Publishers, 1988,
462 Boston Street, Topsfield, MA 01983.

Library of Congress Cataloging-in-Publication Data

Bacon, Josephine, 1942–
   Exotic fruits A–Z.

   Includes index.
   1. Cookery (Fruits)    2. Tropical fruit.    I. Title.
TX811.B33       1988          641.6′4              87–28476
ISBN 0–88162–355–5

Manufactured in Great Britain

# CONTENTS

# INTRODUCTION

The profusion of exotic fruits and vegetables that has been pouring into our shops and markets over the last few years is exciting but also perhaps rather baffling. Exciting because never before has there been such a huge choice of wonderful produce, and baffling because even with the more developed tastes of widely travelled people it is not always obvious what a particular item is, or what should be done with it. This book and its companion, *Exotic Vegetables A-Z,* attempt to solve these pleasant problems by offering all the information you will need in portable, easy-to-use form.

Naturally, different people have different ideas of what 'exotic' means: the strict dictionery definition is simply 'foreign', but in its more general sense it denotes (as far as fruits are concerned) the unusual and, particularly, the tropical. Here the emphasis is on fruits that are likely to be unfamiliar to the average buyer in Europe and North America, and some notes are also given on the buying and preparation of non-native fruits (e.g. pineapple) that may need special care. A few local specialities are also included. 'Fruit' is a broad term too, and I have included a few that strictly speaking are vegetables because they are mainly *used* like fruit. For the same reason, nuts are *not* included here.

One of the biggest puzzles is the multitude of different names by which many of these fruits are known. Physalis, for instance, is variously known as Cape gooseberry, Chinese lantern, goldenberry, ground cherry and Peruvian cherry, depending on where you are and who you are talking to – and it is no relation to either the cherry or the gooseberry! I have adopted the straightforward, if rather drastic, plan of arranging the various fruits in alphabetical order under what I would take to be their most commonly used English names. To make it absolutely clear which fruit is being discussed, this name is followed by the Latin name, and below are listed the most common alternative names. I have included foreign-language names only when they are used in an English-speaking country – but all the alternative names are included in the index at the end of the book, so if the item you want isn't immmediately findable, look there and you will be directed to the right page. Cross references from one entry to another are given in SMALL CAPITALS.

Further identification is helped by line drawings of every single fruit, plus photographs of some of the more interesting-looking ones. The entries themselves are addressed to the questions: Where does it come from? What is it like? When is it ripe, in good condition, available? What can I do with it?

The question of availability is also a ticklish one, for some of these fruits may be very readily found, while others are still very scarce. With improved propagation and cultivation methods and the increasing use of air-freight, more and more fruits are becoming available, however, so watch out for new arrivals at any time. For the same reason, seasonal variations are becoming less of a factor, since as soon as a fruit's season is over in one hemisphere it may well come into season in another, and countries in the southern hemisphere such as Australia, New Zealand and Chile have recently become major exporters to the north, making fruits available at times of the year when they were never seen before.

Often the answer to the question 'What can I do with it?' is of course simply 'Eat it!', but many of these fruits can also be used in cooked dishes from many different cuisines, and where there are interesting possibilities I have included recipes. Again, what may be considered exotic in one place may be fairly commonplace in another, but publishing – like fruit – is rapidly internationalizing and this book will be read by people living on opposite sides of the world. This means that different systems of measurement may be used too, so in the recipes there are three sets: Imperial on the left, then metric and in the right-hand column American. Choose the ones you are happiest with, and stick to one set only – and remember that an Imperial pint contains 20 fluid ounces but an American pint only 16. All spoon measures are level. When cooking fruit for preserves, syrups or jams, try to use copper pans, as the juices can damage other metals, especially aluminium.

Happy hunting, and delicious eating!

# *Acerola*  *Malphigia glabra*

Surinam cherry

This fruit, which bears a strong resemblance to a cherry, grows on a thick bush that is sometimes used as a hedge in tropical and sub-tropical climates. Native to the Caribbean, it has become popular as an ornamental in Florida.

The importance of the acerola has only recently been realized. It is the richest of all fruits in vitamin C, containing as much as 4,000 mg per 100 g of fruit. For this reason, it is now cultivated for medicinal purposes and is a major ingredient in natural vitamin C pharmaceutical products.

The acerola (which, confusingly, is also called the Surinam cherry, like the PITANGA), is too sour to be eaten raw, but gives a good flavour to jams, jellies and preserves, and can be stewed with other fruit.

## SPICED ACEROLA JELLY

| | | |
|---|---|---|
| 1 tablespoon allspice berries | 1 tbsp | 1 tbsp |
| 6 whole cloves | | |
| 1 cinnamon stick | | |
| 16 fl oz dry red wine | 500 ml | 2 cups |
| 2½ lb light brown sugar | 1 kg | 2½ lb |
| 1 teaspoon grated nutmeg | 1 tsp | 1 tsp |
| 1 teaspoon salt | 1 tsp | 1 tsp |
| 4 lb acerolas | 2 kg | 4 lb |
| 1 packet vegetarian gelatine (agar-agar) | 25 g | 1 oz |

Put the whole spices into a small muslin bag or a piece of cheesecloth and tie it securely. Pour the wine into an enamelled or copper preserving pan. Add the bag of spices, the sugar, nutmeg and salt. Cook, stirring constantly with a wooden spoon, until the sugar has dissolved. Then stop stirring and bring the liquid to the boil. Boil briskly for 5 minutes.

Add the acerolas and cook until they are very tender, about 30 minutes. Strain off the liquid and discard the bag of spices, but retain the acerolas.

Combine 8 fl oz (250 ml/1 cup) of the liquid with the gelatine (or follow the manufacturer's instructions if they differ). Return this mixture to the rest of the liquid and stir to dissolve. Leave to cool while you remove the pits from the acerolas. Then stir the acerolas into the liquid. Bottle in sterilized jars. Refrigerate when cold.

This is an interesting substitute for redcurrant jelly or cranberry jelly.

# *Azarole* Crataegus azarolus

This fruit is a member of the hawthorn family, as can be seen by the leaves, which are typical of the species. The fruit is usually yellow, though there are some dark-red varieties. The azarole is native to southern Europe, but it is grown in more northerly climates as an ornamental. Like the rose-hip, it can be made into jams, jellies and preserves.

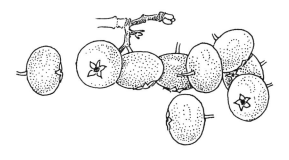

# Babaco *Carica pentagona*

This is the latest of the exotic tropical fruits to be popularized outside its native territory by the resourceful New Zealand farmers. The babaco is an oblong fruit, roughly the same shape as its close relative the PAPAYA. However, the skin is yellow-green when it is ripe, and the flesh is also pale yellow. Unlike the papaya, the fruit has no pips and the skin is edible.

The babaco originates from the Ecuadorian highlands and is a spontaneous cross between the mountain papaya (*Carica pubescens*) and the chamburo (*Carica stipulata*). It was first discovered by European botanists in the 1920s and introduced to New Zealand in 1973. The babaco is now also grown in Guernsey in the Channel Islands for the European market.

Babaco is rich in vitamin C and quite low in sugar. Like papaya it contains an enzyme (papain) which helps the digestion and softens tough meat. The refreshing flavour makes it suitable for combining with meat or fish as well as eating as a sweet. It makes an interesting breakfast substitute for grapefruit.

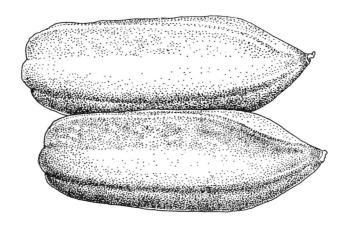

# BABACO COCKTAIL SNACKS

| 2 babacos | | |
|---|---|---|
| 4 oz lean ham | 100 g | ¼ lb |
| 8 oz cottage cheese | 250 g | ½ lb |
| ½ teaspoon powdered sage | ½ tsp | ½ tsp |
| ½ teaspoon dried basil | ½ tsp | ½ tsp |
| ¼ teaspoon cayenne pepper | ¼ tsp | ¼ tsp |
| ¼ teaspoon paprika | ¼ tsp | ¼ tsp |
| parsley sprigs for garnish | | |

Slice the babacos crosswise about ¼ inch (50 mm) thick and arrange them on a dish. Chop the ham roughly and put it in a food processor with the cottage cheese. Add the sage, basil and the cayenne pepper and grind to a purée.

Spread each babaco with some of the purée. Sprinkle lightly with the paprika and spear a parsley sprig on each with a cocktail stick. Serve chilled.

Makes 10–12

# Beach-plum *Prunus Americana sp.*

This wild relative of the cultivated plum is an American native which grows profusely along the sandy shores of New England. The thick white blossoms are to be seen in May. Although the fruits are highly prized, the bushes have a habit of fruiting only sporadically, and efforts to cultivate them commercially have failed. The fruits have a sour flavour which makes them unsuitable for eating raw, but a most delicious jelly is produced from them, which is a New England speciality.

## BEACH-PLUM JAM

| | | |
|---|---:|---:|
| 3 lb beach-plums | 1.5 kg | 3 lb |
| 8 fl oz water | 250 ml | 1 cup |
| 4 lb sugar | 2 kg | 8 cups |

Place the beach-plums and water in a heavy preserving pan. Bring to the boil without stirring. Reduce the heat and simmer for 15 minutes, or until the fruit is soft.

Remove the pan from the heat and leave the contents to cool. When cold seed the plums, retaining all the juice, and trying to keep the fruit whole. Return the plums and juice to the pan. Add the sugar and stir. Boil at medium heat, stirring constantly with a wooden spoon for 15 minutes.

Skim the froth from the jam and spoon the mixture into hot, sterilized jars. Seal and store in a dark, cool, dry place.

Makes about 3 lb (1.5 kg/6 cups)

# Bergamont Orange *Citrus bergamia*

Although the bergamot orange is a hybrid of the SEVILLE ORANGE, it bears a much stronger resemblance to a lemon in that the rind and flesh are lemon yellow. The fruit is round like an orange. It is believed that the bergamot orange's other parent may have been a LIMETTA, because of the presence of bergamot oil in the skin.

The bergamot orange can be used as a souring agent in food, much like the lemon. It is imported mainly from Italy. The name derives from the herb bergamot, to which it is in no way related but which has a similar smell. Bergamot oil is used in perfumery. The blossoms are heavily scented and used as bridal orange blossom. Petitgrain oil, distilled from the young shoots and leaves of the tree, is also an important ingredient in perfumery.

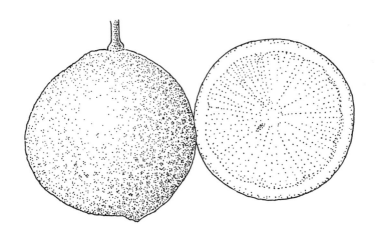

# PORK LOIN WITH BERGAMOT ORANGE

| | | |
|---|---|---|
| 6 bergamot oranges | | |
| 6 orange leaves | | |
| 8 lean slices pork loin | | |
| 1 teaspoon salt | 1 tsp | 1 tsp |
| 1 teaspoon black pepper | 1 tsp | 1 tsp |
| 1 teaspoon chilli powder | 1 tsp | 1 tsp |
| 1 teaspoon Dijon mustard | 1 tsp | 1 tsp |
| ¼ teaspoon tabasco sauce | ¼ tsp | ¼ tsp |
| ¼ teaspoon ground cinnamon | ¼ tsp | ¼ tsp |
| 1 bunch watercress | | |

Grate the rind and squeeze the juice of four of the oranges, discarding the seeds. Slice the remaining oranges crosswise into ½ inch (1 cm) slices, discarding the first and last slices. Wash the leaves and chop them roughly to release the oils.

Sprinkle the meat with the salt, pepper and chilli powder. Put them in a non-stick pan and brown them on both sides. Mix the mustard, tabasco sauce and cinnamon with the grated rind and juice and the crushed leaves, then pour the mixture over the meat.

Cover the pan, reduce the heat and simmer for 20 minutes. Remove the meat from the pan and arrange the slices on a serving dish to keep them warm. Increase the heat and cook the sauce for about 5 minutes to thicken it.

Place a slice of orange on top of each slice of meat and pour the sauce over the top. Garnish with watercress.

4 servings

# *Bilberry* *Vaccinium myrtillus*

Blaeberry (Scotland)
Whortleberry

This close relative of the blueberry (*Vaccinium corymbosum* or *V. angustifolium*) produces single or pairs of berries on the bush, instead of the clusters produced by the blueberry, which is native to North America. The bilberry grows all over northern Europe, and is very popular in Sweden. The round berries are deep purple with a slight bloom on the skin. They are fairly acid, so they are usually cooked and made into pie fillings and other desserts, like the American blueberry.

New cultivated varieties are as large as ½ inch (2 cm) across and sweeter in flavour. They are in season from June to September.

Here is a typical Scandinavian bilberry recipe:

## SWEDISH BERRY CREAM

| | | |
|---|---|---|
| 1 lb fresh bilberries | 500 g | 1 lb |
| 16 fl oz plus 4 tablespoons water | 500 ml | 2 cups |
| 2 tablespoons sugar | 2 tbsp | 2 tbsp |
| 3 tablespoons potato starch or cornflour (cornstarch) | 3 tbsp | 3 tbsp |
| 3 tablespoons water | 3 tbsp | 3 tbsp |
| extra bilberries and whipped cream for decoration | | |

Put the berries into a large saucepan, and add 16 fl oz (500 ml/2 cups) of the water and sugar. Bring to the boil over low heat, then simmer for 3 minutes.

Stir the rest of the water with the potato starch or cornflour and mix to a smooth paste. Stir this paste into the berries and keep stirring until the mixture is smooth, using a wooden spoon and trying not to crush the berries. Bring to the boil and simmer for 3 minutes. Cool and chill. Serve cold in sundae glasses with a few bilberries and whipped cream for decoration.

4 servings

NOTE   Blueberries can be substituted in any bilberry recipe and vice versa.

# Calamondin *Citrus madurensis*

Calamansi
Philippine lime

This fruit, which is grown as an ornamental in Florida and California, occupies a central position in Filipino cooking, because it provides the acid flavour that is lacking in citrus varieties grown in southeast Asia. The climate precludes the growing of lemons, which do not like the excessive heat and year-round rainfall, nor do limes do as well here as elsewhere.

The calamondin looks like a small tangerine, with several seeds. The skin and flesh are deep orange in colour, but the flavour is somewhere between that of a lime and a lemon. The Filipinos squeeze calamondin juice over food as they eat it, or incorporate it into dishes. It is available wherever there are large colonies of emigrant Filipinos; if you cannot get it, substitute a mixture of lime and lemon juice.

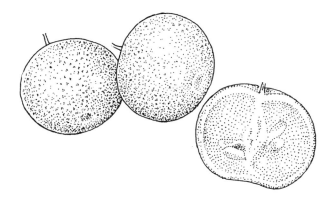

## FILIPINO FISH STEW WITH CALAMONDIN SAUCE

| | | |
|---|---|---|
| 1 lb firm white fish fillets | 500 g | 1 lb |
| 1 teaspoon salt | 1 tsp | 1 tsp |
| 1 tablespoon cooking oil | 1 tbsp | 1 tbsp |
| 1 inch ginger root, chopped | 2.5 cm | 1 in |
| 1 small onion, sliced | | |
| 10 black peppercorns, crushed | | |
| 1 leek, trimmed and sliced lengthwise into strips | | |
| 1 bunch Chinese leaves, cut into 3-in pieces | 7 cm | 3 in |
| juice of 2 calamondins | | |

## CALAMONDIN SAUCE

| | | |
|---|---|---|
| 1 teaspoon cooking oil | 1 tsp | 1 tsp |
| 1 garlic clove, finely chopped | | |
| 1 onion, finely chopped | | |
| 2 tomatoes, peeled and chopped | | |
| 2 tablespoons miso (soya bean paste)* | 2 tbsp | 2 tbsp |
| juice of 2 calamondins | | |
| ½ teaspoon black pepper | ½ tsp | ½ tsp |

Season the fish fillets with salt. Heat the oil in a saucepan with a lid, and sauté the ginger and onion. When the onion turns transparent, add 24 fl oz (750 ml/3 cups) water and the peppercorns. Bring the liquid to the boil. Add the fish and bring back to the boil. Add the rest of the ingredients and cook for 5 minutes. Keep this hot while you make the sauce.

To make the sauce, heat the oil in a frying-pan (skillet) and sauté the garlic and onion until the onion is transparent. Add the tomatoes and simmer until soft. Add the miso. Stir with a fork to make a smooth sauce. Add the calamondin juice and pepper and bring to the boil. Serve with the fish.

* Available from health food stores

# *Carambola* *Averrhoea carambola*

Barbadine
Granadilla (West Indies)
Star fruit

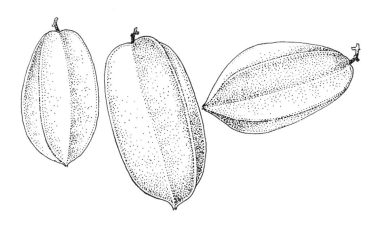

This distinctive tropical fruit probably originated in the Malay archipelago, but is now grown all over the tropics and in sub-tropical climates where water is abundant, including tropical Africa, Brazil and the state of Florida. It has a brilliant-yellow translucent skin, and when cut in cross-section it resembles a star. It has seeds running through one of the 'points' of the star along its length. The skin of the fruit is rather tough, but the flesh is wonderfully fragrant and juicy. It is in season in summer in the various producer countries.

The Latin name for this family comes from Averroes, the famous Moorish philosopher, doctor and astronomer, who lived in Cordoba, Spain, in the 12th century.

The carambola is best eaten raw, occasionally without the skin, in fruit salads, or with cream or cottage cheese. It makes an attractive *nouvelle cuisine*-style garnish for lean red meat or chicken, more unusual than the hackneyed KIWI FRUIT. The carambola is rich in vitamin C. In countries where it grows plentifully, it is used to clean copper and brassware!

A relative of the carambola, the **bilimbi** (*Averrhoea bilimbi*) closely resembles the carambola but is always green in colour, hence its alternative name: the cucumber tree fruit. It is always eaten cooked, in a sugar syrup as a dessert or preserve, or pickled in chutneys. The same treatment is meted out out to unripe carambolas in southern India and China.

## FIVE FINGER DRINK

| | | |
|---|---|---|
| *6 large carambolas* | | |
| *2 pints water* | *1.2 l* | *5 cups* |
| *6 tablespoons sugar* | *6 tbsp* | *6 tbsp* |
| *4 cloves* | | |

Cut or slice the carambolas, retaining all the juice. Place the fruit into a glass pitcher. Add the water and sugar. Add the cloves and cover the pitcher with a cloth. Leave it in a cool place for three days. Strain and serve iced.

Makes 2 pints (1.2 l/5 cups)

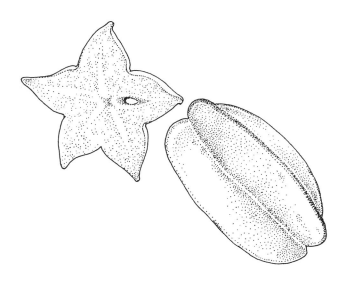

# Carob  *Ceratonia siliqua*

Locust bean
St. John's bread

The beautiful evergreen carob tree is grown as an ornamental outside its native environment, the Middle East. The fruit of the carob tree consists of large pods about 8 inches (20 cm) long, which ripen from green to brown. The pods contain a row of hard, brown seeds, which are discarded for eating purposes. These seeds are so uniform in size and weight that Arab merchants used to use them as measures for weighing precious metals and jewels, and the word 'carat' derives from this.

The two alternative names for the carob come from the story that when John the Baptist went into the wilderness and lived on 'locusts and wild honey', the 'locusts' were believed to be carob pods (though, in fact, in parts of the Middle East, the insects are still a delicacy).

In the Middle East, the sweet pods are chewed by children, and used as animal fodder (donkeys are particularly fond of them). In recent times, carob has been used as a healthier alternative to both chocolate and coffee, since it contains no caffeine or oxalic acid, and has half the fat of cocoa. It is also made into a gum which is used as a binding agent in processed cheeses. Outside the Middle East, whole or broken pods are occasionally available in health food stores. They are tough to chew on, but deliciously sweet (30–50 per cent sugar), and are a pleasant alternative to chewing gum. The beans ripen in late summer and autumn, but keep well, so they are available for most of the year.

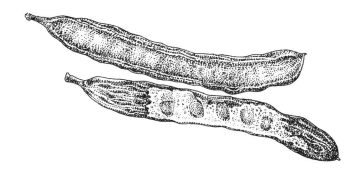

# Cashew apple *Anacardium occidentale*

This large, mango-shaped, red-skinned fruit is part of the cashew tree, though in fact it is not the true fruit, but only the pedicle or base of the flower. The fruit is actually the cashew nut. The scientific genus name refers to the heart-shape of the fruit, and the species name shows that it originates in the western hemisphere. The cashew tree comes from South America and the West Indies, but is now cultivated in many parts of the tropics, especially for the high nutritional value of the nuts.

The cashew apple has a pleasantly sour taste, and is used in the West Indies and Brazil for making preserves and drinks. It is very rich in phosphorus and vitamins A and C. Surprisingly, the cashew is related to poison ivy!

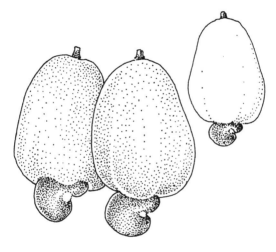

## CASHEW-APPLE DRINK

| | | |
|---|---:|---:|
| 4 large cashew apples | | |
| 1¾ pints water | 1 l | 1 quart |
| 6 tablespoons sugar | 6 tbsp | 6 tbsp |
| 4 whole cloves | | |

Wash and peel the fruit. Put it into a blender or food processor and grind it, adding half the water little by little. Add the rest of the water to the pulp, plus the sugar and cloves, and leave to stand for one day in a cool place. Serve iced.

6 servings

# Cherimoya *Anona cherimola*

Anona
Chirimoya
Custard apple

The members of the Cherimoya family, known as *Anonaceae*, resemble each other in having a thick, soft skin which is inedible, enclosing a creamy flesh, and large seeds. The best-known member of this family is the cherimoya, which is slightly pear-shaped and has a pale-green skin covered in small lobes. The flesh is cream-coloured and has a subtle, creamy taste. The number of seeds varies, depending on where the fruit has been grown.

All the members of the *Anona* family are native to the Americas. The cherimoya originates on the slopes of the Andes, but is now grown throughout the sub-tropics (including California and Florida), and is exported from India and Israel to Europe. The flavour is delicious. It is a winter fruit.

Fruit of the *Anona* family are richer in niacin, phosphorus and thiamin than most fruits, and contain no sodium. Other members of the same family with a similar flavour to the cherimoya are the **sugar apple** (*A. squamosa*), also known as the **sweetsop**, the SOURSOP (which we examine later in the book), and the **ilama** (*A. diversifolia*).

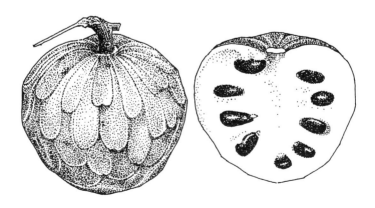

The fruit of all these varietes are ripe when the skin is slightly soft to the touch. They can be bought unripe and should be allowed to ripen at room temperature, in a paper bag. Discoloration of the pale-green skin does not mean the fruit has gone bad. Chill it before eating. The cherimoya is delicious eaten raw, alone or in a fruit salad, but can also be incorporated into desserts. Here is an example:

## CHERIMOYA JELLY DESSERT

| | | |
|---|---|---|
| 1 packet lime jelly (jello) | | |
| 2 small or one large cherimoya | | |
| ½ avocado | | |
| 1 dessert apple | | |
| 2 tablespoons fresh lime or lemon juice | 2 tbsp | 2 tbsp |
| 8 fl oz ginger ale | 250 ml | 1 cup |
| lime cream (p.65) | | |

Make the jelly (jello) according to manufacturer's instructions, but using only half the water. Pour the dissolved jelly (jello) into a bowl and leave to cool to room temperature. While it is cooling, peel the cherimoya(s) and remove all the seeds. Cut into pieces. Cut the avocado and unpeeled apple into cubes. Sprinkle them with the lime or lemon juice.

Pour the ginger ale into the dissolved jelly. Chill, stirring occasionally, until the mixture is the consistency of unbeaten egg white. Then stir in the chunks of fruit. Pour into a 2-quart jelly (jello) mould. Chill until set.

Serve with lime cream.

# Citron *Citrus medica*

The citron is frequently confused with other citrus fruits — and even with completely unrelated fruits — due to the similarity of names. It is the oldest citrus fruit known to the western world, being introduced to Europe in about 500 BC. It is now thought to have originated in Hadramaut, the only well-watered, mountainous region of the Arabian peninsula.

The citron varies enormously in shape and size. The **etrog** variety, which is used in the Jewish festival of Tabernacles (Sukkot) is cultivated in Corfu and Israel in as small a form as possible for convenient export. The favourite examples of the fruit still bear a withered pistil on one end, known as the 'nipple'. On the other hand, the **Corsican** and **diamante** varieties cultivated in France, Corsica and Italy to be made into candied peel and, in Corsica, the liqueur called *cédratine* are twice as big as grapefruits. Both these varieties roughly resemble a very knobbly lemon. Other citrons are ovoid, with deep, widely spaced furrows, so that they look like American footballs. All have a very thick pith and a small area of flesh which is even sourer than a lemon. A very odd variety of the citron, cultivated in China and Japan, is known as **the hand of Buddha**. It is indeed hand-shaped, the long, tapering 'fingers' being segments which have

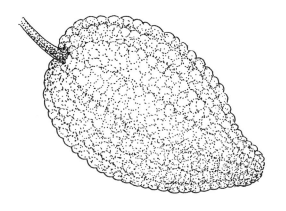

separated on the outside rather than the inside of the plant. It is not eaten, but used purely for perfumery; the strong perfume of all varieties of citron make them suitable for this purpose.

The thick pith of the citron makes it the best fruit for candying. Pumpkin (*citrouille*) is sometimes candied and passed off as citron, and there is also a type of melon called the citron-melon which is confused with the citron. In Mexico, the **cushion cactus** (*Echinocactus grandis*) is candied, and in its candied form is called *acitrón*, no doubt because originally it was a substitute for real candied citron. Nowadays, candied citron is imported into Central and North America from Puerto Rico, where the large rounded varieties of citron grow well.

If you are fortunate enough to find fresh citrons, which are at their best from September to March, they can be either preserved in a sugar syrup or candied. To candy citrons, cut the fruit in half or quarters and soak it in a salt solution for a week, to break down the fibres and ensure that the syrup will permeate the flesh. Then rinse the cut halves well and cook them in a strong syrup until they have absorbed most of the liquid.

# Cloudberry *Rubus chamaemorus*

Baked apple berry (Canada)

The cloudberry looks like a small golden blackberry and is, in fact, a member of the same family. It grows in northern Europe, especially in Finland and Scandinavia, and Canada, flourishing within the Arctic Circle. It ripens in late summer to autumn.

The berries are incorporated into pies and puddings like blackberries, and the Finns produce a cloudberry liqueur.

The Canadian name 'baked apple berry' refers to the flavour of the cooked fruit.

# CLOUDBERRY MERINGUE CAKE

| | | |
|---|---|---|
| 8 oz shelled hazelnuts | 250 g | ½ lb |
| 8 oz castor (superfine) sugar | 250 g | 1 cup |
| 4 egg whites | | |
| 1 lb cloudberries | 500 g | 1 lb |
| 12 fl oz whipping cream | 350 ml | 1½ cups |
| 2 tablespoons sugar | 2 tbsp | 2 tbsp |
| 1 tablespoon cloudberry liqueur | 1 tbsp | 1 tbsp |

Line the bottom and sides of a 9-inch (22.5 cm) loose-bottomed cake tin (pan) or springform tin (pan) with non-stick (silicon) baking paper. Preheat the oven to 350 °F (180 °C/Gas Mark 4).

Put the hazelnuts into a dry non-stick frying-pan (skillet) and cook, stirring frequently, until they begin to give off a nutty aroma. Remove them from the pan and rub them in a kitchen towel to remove the brown outer skins (unless you have bought them ready-skinned). Grind them in a food processor.

Whip the egg whites into soft peaks, then whip in the castor (superfine) sugar and beat until stiff. Fold this mixture into the ground nuts.

Spread the mixture over the bottom of the tin (pan) and bake for 25 minutes or until lightly browned. Carefully remove the sides of the tin (pan) and place the meringue on the cake tin (pan) base on a wire rack. Leave to cool.

Whip the cream with the sugar and liqueur until stiff. Pile the berries on to the meringue base, reserving a few for garnish, and cover with whipped cream. Garnish with the remaining berries.

6 servings

# Coconut *Cocos nucifera*

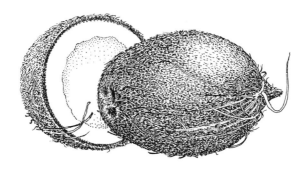

The coconut is one of the most familiar 'exotic' fruits in northern countries. Its excellent keeping qualities made it possible to ship coconuts even before refrigeration was invented.

Coconut palms enjoy the environment of the shoreline, being able to tolerate the high level of salt in the sandy soil. They will not grow in hilly regions, and are confined exclusively to the tropics. They grow throughout the equatorial region, but probably originated in the Old World.

The coconut itself consists of an outer coating or husk, which is brownish-green in colour when ripe. Inside this husk is a mass of matted fibres called copra surrounding the hard shell of the nut. The centre of the nut, beneath the layer of hard white flesh, is hollow but contains a liquid which some people call coconut milk, but which is more correctly referred to as coconut water.

Nowadays, one can buy 'jelly coconuts', unripe coconuts still in the husk, in West Indian shops. They can be cut open with a large kitchen knife. The flesh is soft and jelly-like, and considered a great delicacy.

When choosing ripe coconuts, examine the fibrous tuft, and look at the 'eyes', which must be free of any signs of mould, and shake the coconut to listen to the water swirling around. The water dries up as the coconut gets older; the less water, the less juicy the coconut flesh, so choose one with plenty of water.

To prepare a ripe coconut for eating, first cut off the fibrous tuft on the top of the coconut. This will reveal the three 'eyes'. Pierce the eyes with a thick skewer or hand-drill, and pour the water into a bowl. The water is a delicious addition to long drinks (especially those with rum), and to desserts. Try it with pineapple chunks.

To open the hard shell, tap it with a hammer or the back of a meat cleaver all around the middle, about one-third of the way from the end opposite the tuft. The coconut should crack along a natural seam. It can then be smashed further to facilitate removal of the flesh. Pull the flesh away from the shell, and carefully cut away the thin brown skin, which would otherwise discolour the flesh when it is grated. Then chop or grate the flesh.

Coconut milk, made from the fresh grated flesh, is an important ingredient in a wide variety of tropical dishes (including several recipes in this book). Although coconut milk can be bought canned, sweetened or unsweetened, it tastes far better when home-made.

To make coconut milk, grate coconut flesh into a large bowl. Measure it in an 8-fl-oz (250 ml) American standard measuring cup. For every two cups of coconut pour one cup of boiling water over the coconut. Leave to stand at room temperature for 30 minutes. Strain the liquid through cheesecloth, wringing the cloth to get all the liquid out. Refrigerate until required. It keeps for about as long as fresh milk.

This process can be repeated, using the same grated coconut flesh. Sometimes the first pressing is called coconut cream, and the subsequent pressings are called coconut milk. Never boil coconut milk or it will curdle.

Here is a simple recipe using coconut cream or milk as a dessert:

### HAWAIIAN COCONUT PUDDING

| | | |
|---|---|---|
| *16 fl oz coconut cream or milk* | *500 ml* | *2 cups* |
| *3 tablespoons sugar* | *3 tbsp* | *3 tbsp* |
| *3 tablespoons cornflour (cornstarch)* | *3 tbsp* | *3 tbsp* |
| *¼ teaspoon salt* | *¼ tsp* | *¼ tsp* |
| *3 tablespoons white rum* | *3 tbsp* | *3 tbsp* |
| *Pineapple rings and angelica stems to decorate* | | |

Pour 4 fl oz (125 ml/½ cup) of the coconut milk into a bowl and combine it with the sugar and cornflour (cornstarch). Add the salt. Heat the rest of the coconut milk, and when it is hot, stir in the cornflour (cornstarch) mixture. Stir constantly, over low heat, until the mixture is thickened, but do not let it boil. Remove the pan from the heat and add the rum.

Pour the mixture into a glass serving dish. Cool to room temperature, then refrigerate until cold. Decorate with pineapple rings and angelica stems before serving.

4–6 servings

# *Cornel*  Cornus mas

Cornelian cherry
Dog cherry
Siberian cherry
Tartar cherry

This member of the dogwood family grows on a small tree, whose yellow flowers bloom in February, even before the leaves unfold. The fruit is red and ovoid, something like a rose-hip. The tree is grown extensively in southern Europe as an ornamental.

Cornels are an important ingredient in Russian cooking, especially in central Asia, where they are used much like the more familiar sour cherry. They give a sweet-and-sour flavour to meat dishes and desserts, and can be bought outside the USSR in semi-dried form.

In France, cornels are pickled like olives or made into preserves.

## CORNEL CAKE

| | | |
|---|---|---|
| 4 oz cornels, fresh or dried | 100 g | ¼ lb |
| 2 tablespoons cherry brandy or kirsch | 2 tbsp | 2 tbsp |
| 6 oz butter | 175 g | ¾ cup |
| 6 oz sugar | 175 g | ¾ cup |
| 8 oz flour | 250 g | 1 cup |
| 2 teaspoons baking powder | 2 tsp | 2 tsp |
| ¼ teaspoon salt | ¼ tsp | ¼ tsp |
| 3 eggs, well beaten | 3 | 3 |
| 2 oz ground almonds | 50 g | ¼ cup |

Line an 8-inch (20 cm) cake tin (pan) with non-stick baking paper. Preheat the oven to 350 °F (180 °C/Gas Mark 4). Remove the pits from the cornels and soak them in the liqueur. Cream the butter and sugar until smooth. Sift the flour, baking powder and salt into a bowl. Beat in the butter and sugar mixture. Add the eggs and ground almonds. Drain the cornels, adding the liquid to the mixture. Beat it into a smooth dough. Roll the cornels in flour and incorporate them evenly into the dough.

Bake the cake for 1½ hours or until a toothpick inserted in the centre comes out clean. Cool on a wire rack. Ice when cold.

8 servings

# $\mathcal{D}$ate *Phoenix dactylifera*

Dates have been cultivated for so long that the wild fruit has disappeared. Fresh dates when picked do not ripen evenly or keep well, and for that reason, they are usually smoked to preserve them for export. The smoking process hastens uniform ripening and makes them sweeter.

Recently, the Israelis have started exporting several varieties of dates in their raw state, no doubt for the sake of novelty, but they do not export well in the fresh state, and tend to be dry and fibrous. Iraqi dates are sold in blocks for use in cooking, and Tunisian dates are exported in boxes, with the central stem still in place.

Dates ripen in the autumn. They grow in huge bunches on the date palm, at the end of long, narrow stalks, and are to be found in all the arid sub-tropical regions of the Middle East, Africa and India. There is also a major date-growing centre around Indio, in southern California.

There are over 350 varieties of dates, including the light-brown **Deglet Noor**, the most popular of all the varieties. Dates are classified as 'soft' or 'wet' and 'dry'.

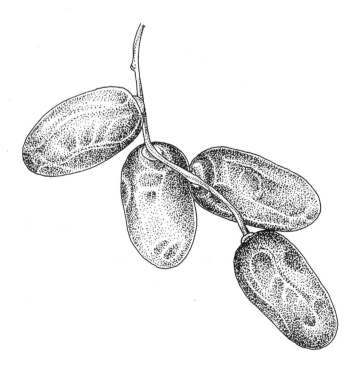

The yellow dates are usually considered to be soft, and the red, dry. Dry dates are not often seen outside the Middle East and North Africa, though one dry variety, the **Thuri**, is grown in southern California. The exception is the large, red date of Mesopotamia, now also grown in Israel, which is the best of the wet varieties, is the **Halawi** (the word for 'sweet' in Arabic).

Dates are often eaten raw, but they can also be incorporated into cakes and cookies, and eaten, as the Arabs do, chopped with sour milk and yoghurt. The date growers around Indio sell dry date granules, which can be whipped with milk into date milk shakes. The Arabs also fry dates in butter, and eat them with meat. Dates stuffed with almonds make attractive candies. The grooved seed can be extracted from one end with an olive pitter, or the date can be slit lengthwise.

## DATE CAKE

| | | |
|---|---|---|
| *2 large eggs* | | |
| *4 oz sugar* | *100 g* | *⅓ cup* |
| *¼ teaspoon salt* | *¼ tsp* | *¼ tsp* |
| *6 oz breadcrumbs* | *175 g* | *¾ cup* |
| *1 teaspoon baking powder* | *1 tsp* | *1 tsp* |
| *1 teaspoon cinnamon* | *1 tsp* | *1 tsp* |
| *1 lb chopped dates* | *500 g* | *2 cups* |
| *6 oz chopped nuts (almonds or walnuts)* | *175 g* | *¾ cup* |
| *1 teaspoon vanilla essence (extract)* | *1 tsp* | *1 tsp* |

Grease an 8-inch (20 cm) baking tin (pan) or cake tin (pan). Preheat the oven to 350 °F (180 °C/Gas Mark 4). Beat the eggs and beat the sugar and salt into them until the mixture turns pale. Add the vanilla. Mix the breadcrumbs with the baking powder and cinnamon. Stir the dates and nuts into the dry mixture. Add this mixture to the egg mixture, beating well. Bake the cake for 45 minutes. Serve hot as a pudding with cream or ice cream, or leave to cool and serve as cake.

8–10 servings

# $\mathcal{D}urian$ *Durio zibethinus*

Civet-cat fruit

This fruit is notorious for its pungent and rather nasty smell, which is in contrast to the sweet, pleasant taste of its flesh. It is highly prized in southeast Asia, and also grows in East Africa. Because of the smell, it is rarely exported in fresh form, but is often seen canned and incorporated into cakes and cookies.

The fruits are large, weighing as much as 10 lb (4.5 kg), and are completely covered in greyish-yellow spines. The flesh is creamy-yellow, and encloses up to six large brown seeds. The flavour is so delicious that it completely makes up for the obnoxious smell. Elephants and tigers are said to be particularly partial to durians, and can detect them from far away thanks to the pervasive odour.

It is advisable to eat fresh durian immediately, as the smell will pervade the refrigerator or anywhere else you keep it, but the pulp can be frozen in a plastic bag and will then be odourless. It can be made into fruit salad or stewed.

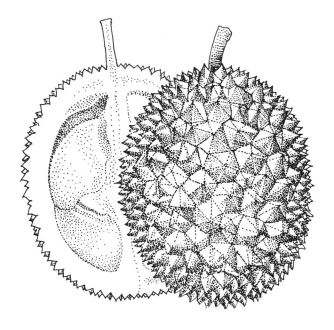

# DURIAN CAKE

| 1 durian | | |
|---|---|---|
| 6 oz butter | 175 g | ¾ cup |
| 5 oz sugar | 150 g | ⅔ cup |
| 4 eggs, separated | 4 | 4 |
| 6 oz flour | 175 g | ¾ cup |

Preheat the oven to 350 °F (180 °C/Gas Mark 4). Remove the flesh from the durian and discard the seeds. Purée the flesh in a food processor. Beat the butter and sugar together until smooth, then beat in the egg yolks. Sift the flour and beat it into the butter mixture. Beat in the durian pulp.

Whip the egg whites into stiff peaks and fold them into the mixture. Bake the cake in a greased, 8-inch (20 cm) cake tin for 1 hour, or until the cake is done when tested with a toothpick.

8–10 servings

*Right: a durian, with one of its interior segments in the foreground. The dessert spoon gives an idea of the durian's large size.*

# *Feijoa*  *Feijoa sellowiana*

Brazilian guava
Guavasteen (West Indies)
Pineapple guava

This little green-skinned fruit, which can be shiny or slightly downy in appearance, has a very similar taste to the GUAVA, though it is only a distant relative, being a plant genus of its own. It is named after the Portuguese botanist who discovered it in Brazil, Dom da Silva Feijoa. Inside the feijoa, the arrangement of black seeds in a red pulp is similar to the central arrangement of the guava's white seeds.

The feijoa is well known in the southwestern United States, but relatively rare in Europe, though it is just being introduced from South America and New Zealand. Feijoas are in season throughout the summer. They can be eaten in fruit salad or stewed, but are best eaten raw on their own. The skin is rather tough and is sometimes discarded.

# $\mathcal{F}ig$ *Ficus carica*

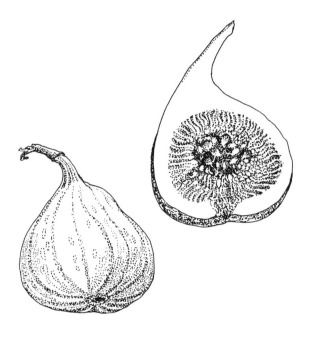

Figs are rounded with a pointed stem end and vary in colour when ripe from bright green to deep purple, depending on the variety. They have been cultivated since prehistoric times, and are extremely useful, since they can be dried for storage. The flavour is sweet and fragrant.

The best known varieties are **black mission**, a greyish-black type developed in Mexico and California; **kadota**, a green fig ripening to pale yellow mainly grown in Greece, Italy and California; and the purple variety grown in Greece, Cyprus and Israel and known as **sari lob**. Other well-known varieties are **brown Turkey**, which has a reddish-brown skin. All figs have a ruby-red juicy flesh, with white protrusions of firmer flesh. They contain either a large number of minute seeds, or none at all, depending on whether or not the fruit has been fertilized.

The fig tree exudes a white latex, which is used in the Middle East to curdle milk for cheeses and yoghurt-making. Fresh figs taste particularly good with the white, salty cheeses of the Middle East, such as feta and halloumi. When choosing figs, make sure they are not shrivelled and that the flesh is whole and unblemished.

Cut them in half before eating and examine them for grubs; there is a white grub which invades figs which strongly resembles the white parts of the fig.

Canned figs are delicious with ice cream or fresh cream, and dried figs can be stewed, or stuffed with nuts and eaten as candy.

## FIG AND SWEET POTATO PUDDING

| | | |
|---|---|---|
| 1 lb sweet potatoes, boiled and peeled | 500 g | 1 lb |
| 8 oz dried figs, finely chopped | 250 g | 1 cup |
| 5 oz dry breadcrumbs | 125 g | ⅔ cup |
| 6 oz seedless raisins | 150 g | ¾ cup |
| 8 oz butter, half of it melted | 250 g | 1 cup |
| 4 oz sugar | 100 g | ½ cup |
| 3 eggs, beaten | 3 | 3 |
| 10 fl oz milk | 300 ml | 1¼ cups |
| 1 teaspoon cinnamon | 1 tsp | 1 tsp |

Grease a deep 8-inch (20 cm) ovenproof dish. Preheat the oven to 350 °F (180 °C/Gas Mark 4). Mash the sweet potatoes or push them through a sieve. Add the figs, breadcrumbs, raisins and the melted butter. Beat in the sugar, eggs and milk.

Turn the mixture into the greased dish. Sprinkle it with the cinnamon and dot with the rest of the butter. Bake for 40 minutes or until golden brown. Serve with custard.

8 servings

*Right: figs in various stages of ripeness, one cut to show the interior of a fresh fig.*

# $\mathcal{F}$ruit salad fruit _Monstera deliciosa_

Ceriman
Monstera Deliciosa

This is the fruit of the Swiss cheese plant, the only plant whose leaves are naturally slashed and contain holes. The fruit is so called because it is claimed that if you cut off a piece, eat it and leave the rest in the refrigerator for a while before cutting off another, the next piece will taste quite different from the first: something of a chameleon among fruits.

The Swiss cheese plant does not usually bear fruit when cultivated in hothouses or indoors in cooler climates. It is a native of tropical America, and the cone-like fruit is popular there, and it has recently been grown commercially in Australia. It remains to be seen whether the fruit will be marketed commercially in northern Europe and the U.S.A.

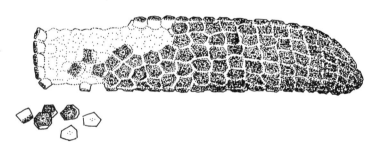

# *Golden apple* Spondias cytheria

The golden apple is, of course, not an apple at all but a yellow, egg-shaped fruit with a large single seed, rather like that of an avocado. Although it is very popular in the West Indies, it actually originates from Polynesia. The fruits are occasionally exported, and are in season in late autumn and winter.

Many of the fruits have thick fibres among the pulp, which should be removed during preparation. Golden apples are made into jams and drinks.

# GOLDEN APPLE CHUTNEY

| | | |
|---|---:|---:|
| 2 lb golden apples | 1 kg | 2 lb |
| 16 fl oz vinegar | 500 ml | 2 cups |
| 2 onions, chopped | 2 | 2 |
| 4 oz raisins | 100 g | ½ cup |
| 12 oz brown sugar | 300 g | 1½ cups |
| 1 teaspoon salt | 1 tsp | 1 tsp |
| 1 teaspoon ground ginger | 1 tsp | 1 tsp |
| ½ teaspoon ground cloves | ½ tsp | ½ tsp |

Put the golden apples into a preserving pan or other non-aluminium pan. Add the vinegar and bring to the boil. Reduce the heat and simmer for 10 minutes. Strain, and reserve half the liquid, discarding the rest.

Remove the seeds from the golden apples and discard as many thick fibres as possible. Return the golden apples and liquid to the pan and add the onions. Boil for 5 minutes, stirring constantly. Add the rest of the ingredients and cook, stirring, until the mixture thickens — about 15 minutes. Bottle and seal.

Makes about 2¼ pints (1.5 litres/6 cups)

*Right: kiwanos, kiwis and kumquats (see pages 58-62). The orange, spiky kiwanos are at the top, with the kiwi fruits below surrounded by kumquats.*

# _Governor plum_ Flacourtia indica

Botoko plum
Ramontochi

This plum-shaped fruit is of Indian origin, as its name implies, though it is also to be found in tropical Africa. However, it is most popular in the West Indies, and can be found in season in West Indian markets in Europe and the U.S.A.

The small purple fruits are formed in late summer and are about the size of marbles, so that they are the tropical equivalent of damsons. Like damsons, they are full of pectin and make an excellent jelly. Governor plums require about 10 oz (300 g/1½ cups) sugar to each 8 fl oz (250 ml/1 cup) of juice. They are good cooked with fish, and are rich in calcium and vitamin C.

# $\mathcal{G}uava$ Psidium guajave

The guava is a native of South America, but is now grown in southeast Asia, Florida and Australia. The fruit has a shiny yellow skin, which is irregularly pitted and furrowed, and it is the size and shape of an apple. Like apples, guavas range in size from that of a small egg to that of a large orange. The flesh is firm but smooth, though sometimes grainy. The white seeds are held in the centre of the fruit in a jelly-like pulp. The seeds are sometimes soft enough to eat with the raw fruit, but in the larger fruits they are often hard and should be discarded.

Guavas have a strong and pervasive odour, and should be well wrapped when refrigerated raw. For this reason, it is inadvisable to display them in the fruit bowl, as the smell, which not everyone finds pleasant, will pervade the house. The odour disappears on cooking.

Guavas are in season in the summer. They are very rich in vitamin C and, since that vitamin deteriorates quickly, are best eaten soon after they are bought. Do not buy shrivelled or pitted fruits, and make sure they are firm and not bruised.

Guavas are often considered to be best eaten raw. They are delicious with cottage cheese or in a tropical fruit salad with pineapple, mango and papaya.

In South America and the West Indies, they are often cooked like apples, and are also made into a paste-like preserve called *guanabana* (not to be confused with the fruit of the same name). In the Philippines, guavas are mixed with meat and fish in stews.

A related variety, with a pink skin and flesh, the **strawberry guava** (*P. cattleianum*), is used in the same way, and is often sold canned. It is also puréed and used to add a delicate pink colour to long fruit-flavoured drinks, particularly in Brazil.

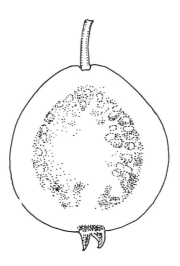

*Left: guavas are at the bottom of the picture, with (above, left) papayas and (right) mango.*

# GUAVAS STUFFED WITH COCONUT CREAM

| | | |
|---|---|---|
| 6 guavas | | |
| 8 oz sugar | 250 g | 1 cup |
| 4 oz unsweetened, shredded coconut | 125 g | ½ cup |
| 6 tablespoons unsweetened coconut milk | 6 tbsp | 6 tbsp |
| 2 egg yolks | 2 | 2 |
| ½ teaspoon ground cloves | ½ tsp | ½ tsp |
| ½ teaspoon rosewater | ½ tsp | ½ tsp |
| 8 fl oz orange juice | 250 g | 1 cup |
| Grated rind of 1 orange | | |

Slice the guavas in half and put them in a pan with half the sugar. Add 16 fl oz (500 ml/2 cups) water and bring to the boil. Cover the pan and simmer until the guavas are tender, about 20 minutes. Remove them, reserving the syrup. Leave them to cool, then scoop out the seeds and some of the flesh, leaving the guava shells with a thick layer of flesh.

Measure one cup of the guava cooking liquid into a heavy-based pan. Add the rest of the sugar and bring to the boil. Boil briskly for 10 minutes. Add the coconut and cook until the coconut is transparent, about 5 minutes. Remove from the heat and allow to cool slightly.

Combine the coconut milk with the egg yolks, spices and rosewater. Add this mixture to the coconut, and return to the heat. Cook, stirring with a wooden spoon, until the mixture forms a thick cream. Remove it from the heat and use it to fill the guava shells, spreading it over the surface of the fruits.

Pour the orange juice into a shallow serving dish and arrange the guavas, filled side up, in the juice. Sprinkle with the grated zest. Chill before serving.

6 servings

# Hog-plum *Spondias mombin*

This plum-like fruit is native to the Caribbean and the southern states of the United States. It is not grown commercially but is very common, and so is often used to make jam or jelly. Hog-plum jelly requires 8 oz (250 g/1 cup) sugar to be added to 8 fl oz (250 ml/1 cup) fruit juice.

The hog-plum is only about 1 inch (2.5 cm) in length, and has a large seed set in a thin orange-yellow pulp and skin. It is extremely rich in vitamin C, and is a good source of calcium and phosphorus.

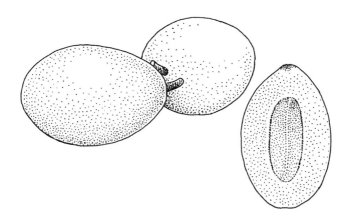

# *Jackfruit* *Artocarpus heterophyllus*

Jakfruit

This enormous fruit, the largest known, can be up to 3
feet (1 metre) long and 20 inches (500 cm) across! The
heaviest fruits can weigh 90 lb (40 kg)! However, smaller
specimens weighing about 4 lb (2 kg) are exported. In
fact, the jackfruit is really a collection of fruits which fuse
together. (The same is true of a related fruit, the FIG.) A
close relative of the breadfruit (which is classified as a
vegetable and to be found in the companion volume to
the present book), the jackfruit has a similar hard, greenish-yellow skin, completely covered with pointed warts.
The large seeds are enclosed in fleshy sacs called 'bulbs'.
The seeds of the jackfruit can be roasted, and then they
taste similar to chestnuts. They are very rich in calcium

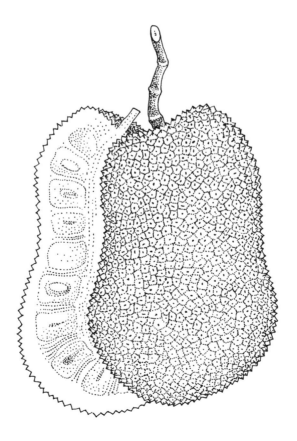

and contain 12 per cent protein, but the fruit itself is not very rich in vitamins, though it contains carotene. The flesh can be eaten raw or cooked, and eaten with meat or as a dessert.

Jackfruit is of Indian origin, but it is also grown in East Africa. The tree flowers and fruits almost all year round, but is at its best in mid-summer to early autumn. Jackfruit was brought to South America by Europeans and now grows there as well.

Jackfruits are made into preserves and used in fruit salads and candies. Choose firm, undamaged fruits. Some varieties are rather dry and flavourless, and unfortunately it is very hard to tell from the external appearance whether the fruit will be sweet and juicy, though a strong smell is a good indication that it will.

## JACKFRUIT PRESERVE

| | | |
|---|---:|---:|
| 2½ lb jackfruit | 1.25 kg | 2½ lb |
| 1 pineapple | 1 | 1 |
| 2½ lb preserving sugar | 1.25 kg | 2½ lb |
| 4 tablespoons liquid glucose | 4 tbsp | 4 tbsp |
| 2 tablespoons fruit pectin | 2 tbsp | 2 tbsp |

Peel the jackfruit and remove and discard the seeds. Peel the pineapple, core it and remove the eyes and cut it into chunks.

Put the sugar and liquid glucose into a preserving pan and add 8 fl oz (250 ml/1 cup) water. Bring to the boil without stirring and boil 5 minutes. Add the fruits and reduce the heat. Add the pectin and simmer, stirring occasionally to prevent sticking, for 45 minutes or until the mixture jells when dropped from a teaspoon on to a cold saucer. Cool and bottle in sterilized jars.

Makes 2 lb (1 kg)

# Jamaican sorrel *Hibiscus sabdariffa*

Red sorrel
Roselle
Shrub
Sorrel

This member of the hibiscus family is native to West Africa. It was imported into the Americas at the time of the slave trade, and is popular in the West Indies, Brazil and Mexico. The Jamaican sorrel fruit and flowers are used to make jams and drinks. In the late eighteenth and early nineteenth centuries, when many new drinks were invented, it was popular in the southern United States as the drink called shrub, and as a flavouring in other long drinks. Its attractive ruby-red colour makes it a good substitute for POMEGRANATE juice (grenadine).

Outside its native soil, Jamaican sorrel is usually sold dried or as a syrup, though fresh sorrel can be bought around Christmas-time. The fruit is enclosed in the fleshy, ruby-red flower petals. The leaves and stalks have a pleasantly acid flavour, like that of sorrel (hence the plant's name, though it is a member of a completely different family) and it can be used in salads and seasonings. Jamaican sorrel is rich in vitamin C.

To make a drink from Jamaican sorrel, steep the flowers and fruits in boiling water for 1 hour. Strain, and add sugar to taste. Chill until ice cold. In Barbados, dark rum is added to the drink.

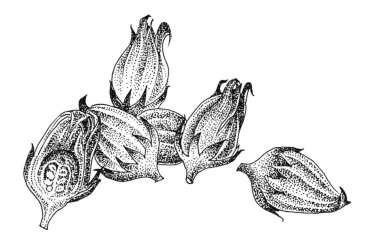

# SORREL JELLY

| 2½ lb fresh Jamaican sorrel (about 1 lb dry sorrel) | 1.25 kg | 2½ lb |
|---|---|---|
| about 2½ lb sugar | 1.25 kg | 2½ lb |
| water | | |

Wash the Jamaican sorrel carefully, and put it in a large china bowl. Add boiling water to cover and stand overnight at room temperature. The next morning, pour the mixture into a preserving pan. Boil, stirring and squashing the fruit with a wooden spoon, until the mixture forms a pulp.

Strain the pulp through a jelly bag or cheesecloth, allowing the juice to drip through slowly. Measure the juice in a measuring cup and measure out the same volume of sugar. Pour juice and sugar into a preserving pan and simmer for 20 minutes, or until the jelly sets when a teaspoonful is dropped on to a cold saucer. Bottle in sterilized jars.

Makes about 2½ lb (1.25 kg)

# *Jujube* Ziziphus jujuba

Chinese date
Jujula
June plum (Caribbean)

These olive green fruits, about the size of a hen's egg, have a single stone firmly embedded in the flesh. The flesh is firm, like that of an unripe pear, and is sweet-sour, with a slight flavour of turpentine. It contains tough fibres which need slow cooking, though the fruit can be eaten raw.

Jujubes are grown all over Asia, especially in Thailand and India, but are believed to have originated in China. They are also eaten in the Caribbean. Jujubes can be stewed or dried, and candied.

Jujubes keep well in the refrigerator for about three weeks. The **French** or **Italian jujube** is a related variety found in Mediterranean Europe; it is yellower in colour.

# JUJUBE CHUTNEY

| 2 lb jujubes | 1 kg | 2 lb |
|---|---|---|
| 16 fl oz wine vinegar | 500 ml | 2 cups |
| 2 onions, chopped | 2 | 2 |
| 4 oz raisins | 100 g | ½ cup |
| 12 oz brown sugar | 300 g | 1½ cups |
| 1 teaspoon salt | 1 tsp | 1 tsp |
| 1 teaspoon ground cinnamon | 1 tsp | 1 tsp |
| ½ teaspoon ground cloves | ½ tsp | ½ tsp |

Put the jujubes into a preserving pan or other non-aluminium pan. Add the vinegar and bring to the boil. Reduce the heat and simmer for 10 minutes. Strain, and reserve half the liquid, discarding the rest.

Remove the seeds from the jujubes and discard any 'strings'. Return the jujubes and liquid to the pan and add the onions. Boil for 10 minutes stirring constantly. Add the rest of the ingredients and cook, stirring, until the mixture thickens, about 15 minutes. Bottle in sterilized jars and seal.

Makes about 2¼ pints (1.5 litres/6 cups)

# *Kiwano* *Cucumis melo sp.*

African horned melon

This striking member of the cucumber family has been developed for export in New Zealand under the name 'kiwano', to cash in on the popularity of the KIWI FRUIT, which the New Zealanders also renamed for export. The fruit is similar in shape to a large gherkin, but is bright orange with little 'horns', each of which is circled by a darker orange ring which looks just as if it had been painted on with a brush. Inside, the flesh is bright green, and contains numerous seeds, which are eaten with the pulp. The skin is thick and inedible. The flavour resembles a combination of mango and pineapple.

Despite its high price in the shops, the African horned melon grows prolifically, and the fruit is likely to become less expensive, especially if growers in other semi-tropical climates take the hint and start producing their own kiwanos.

The kiwano should be eaten raw, and is an excellent addition to fruit or vegetable salads. It is in season in the New Zealand summer, i.e. about October to March.

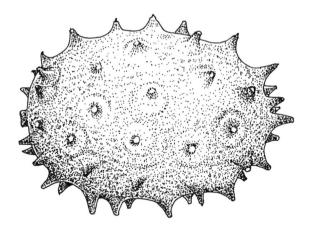

# *Kiwi fruit* *Actinidia sinensis*

Chinese gooseberry

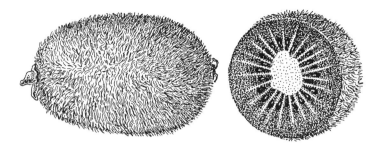

The kiwi fruit, once an expensive rarity outside its native China, has been so heavily popularized by the New Zealand growers, who even popularized a new name for it, that it is now a commonplace that hardly needs description. It is said that the New Zealand exporters originally changed the name to overcome American prejudice, which at the time — the McCarthy era in the 1950s — was rife against anything Chinese!

The fruit grows prolifically on vines, and is now exported not only from New Zealand but also from Kenya, Italy, Israel and California. It combines well with other fruit in a fruit salad, and tastes wonderful in ice cream or with meat. It is so decorative that it is frequently used as a garnish for fish or meat, and is in season all year round. Choose firm, plump fruits that are unblemished. They should 'give' slightly when squeezed if they are ripe, although they can always be ripened in a warm place in a paper bag. When very ripe they can be peeled with the fingers or a sharp knife. If the skin does not come off easily, dip the fruit in boiling water for a few moments.

# COD FILLETS WITH KIWI FRUIT

| | | |
|---|---:|---:|
| *3 kiwi fruit* | | |
| *2 tablespoons butter* | *25 g* | *2 tbsp* |
| *4 cod fillets (about 1½ lb)* | *750 g* | *1½lb* |
| *1 lemon, juice squeezed* | *1* | *1* |
| *½ teaspoon salt* | *½ tsp* | *½ tsp* |

Peel the kiwi fruit. Chop two of the fruits and slice the third. Melt the butter in a shallow pan with a tight-fitting lid. Add the cod fillets and the lemon juice. Cover the pan and cook on very low heat for 15 minutes. Add the chopped kiwi fruit, sprinkling it over the fish, and cook for another 10 minutes.

Arrange the fillets on a serving dish and decorate with the sliced kiwi fruit.

4 servings

# *Kumquat* Fortunella sp.

The kumquat looks like a miniature orange, although botanists have placed it in a different category for various genetic reasons which need not concern us here. To all intents and purposes, it *is* a miniature orange, and it is the only member of the citrus family whose skin is as edible as the flesh. The Chinese name means 'give friend gift', and potted kumquat plants were often exchanged as presents. Although the fruit is miniature, the plant is not naturally a dwarf — in fact, kumquat trees can be among the largest of citrus trees.

The two principal varieties available commercially are both of Japanese origin: the **nagami** (*Fortunella margarita*), an oval kumquat; and the **meiwa** (*F. crassifolia*), which is rounded and slightly larger. Unfortunately, because kumquats are more susceptible than other citrus fruits to the Mediterranean fruit fly, their importation is occasionally banned, so their appearance on the market is sporadic.

The fruits are in season from November to February in the northern hemisphere, and they should always be bright orange and firm to the touch. The most spectacular (but, perhaps, hackneyed) use for kumquats is to decorate the rib ends of a crown roast of lamb. Kumquats are delicious candied, and can be served with ice cream or other plain desserts.

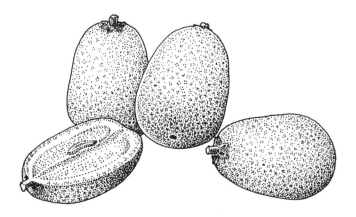

# CANDIED KUMQUATS

| | | |
|---|---:|---:|
| 2½ lb kumquats | 1.25 kg | 2½ lb |
| 4 tablespoons coarse salt | 4 tbsp | 4 tbsp |
| 3 lb sugar | 1.5 kg | 3 lb |
| 8 oz icing (confectioner's) sugar, sifted | 250 g | 1 cup |

Slice the kumquats crosswise into three sections, removing and discarding any seeds. Put them into a preserving pan and add water to cover. Add the salt and leave for 24 hours at room temperature.

Drain the fruit and rinse under cold running water. Put them in a large pan and add water to cover by 2 inches (5 cm). Bring the water to the boil and boil for 15 minutes or until the fruit is tender. Drain thoroughly, discarding the water.

Have ready four wire cake cooling racks, or other racks, covered with wax paper. Put 1 lb (500 g) of the sugar into a preserving pan and add 1¾ pints (1 litre/1 quart) of water. Bring to the boil, stirring only until the sugar dissolves. Add a handful of kumquats and boil them until they are translucent. Remove and drain them on wire racks and replace in the pan with another batch of fruit.

When all the kumquats have been cooked, add the rest of the sugar to the pan and bring it to the boil again, without stirring. Return the kumquats to the pan and boil them until the syrup registers 250 °F (120 °C) on a sugar (candy) thermometer, the firm ball stage.

Carefully remove the kumquats with a skimmer and arrange them on the wire racks again to drain and cool. Sift the icing (confectioner's) sugar over them and leave them to dry in a warm place or in an oven on the coolest setting. Place the candies in airtight tins lined with wax paper and sprinkled with icing (confectioner's) sugar.

Makes about 5 lb (2.25 kg/10 cups)

# *Lime*  *Citrus aurantifolia*

Green lemon

This fragrant, green citrus fruit is cultivated mainly in the tropics, and in those sub-tropical areas where water is plentiful. It originated in India but, unlike the other popular citrus fruits, it is generally not hardy enough to be cultivated in the desert climate of the Mediterranean, though attempts are being made to cultivate it in Israel for export. In Egypt, however, it is commoner than the lemon. Limes are dried in the sun and added to stews and soups in southern Iran and Iraq. The inside of the lime turns black, but the delicious flavour is retained. It is perfectly possible to dry limes in cool climates by putting them on a warm radiator.

Limes were found to grow well in the West Indies, and supplies were shipped to Britain, where they were issued with the rum ration to sailors to prevent the nutritional deficiency known as scurvy. This practice only began in the seventeenth century, though to the disgrace of the British Navy, the curative properties of citrus fruits in

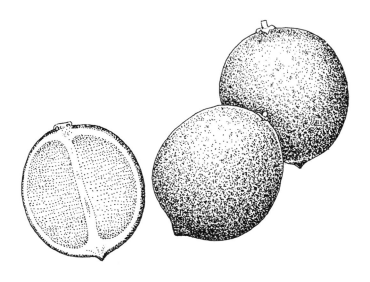

fighting scurvy were known at least 100 years prior to that time. This is how British sailors, and eventually other British emigrants to the United States and Australia, came to be known as 'limeys'. Limes for the navy were stored in warehouses in the London docks in the district which thus became known as Limehouse, and Lime Street in Liverpool is so named for the same reason. In fact, lemons contain twice as much vitamin C (the anti-scorbutic factor) as limes, but lemons grew in the Mediterranean region, where there was a lot of hostility to Britain, and limes grew in British possessions.

Limes eventually ripen to yellow skins, but are usually picked when green. The flesh always remains green. The fruit is so fragrant that the oil from the skin is used in perfumery, and lime juice is sweeter than lemon juice. Of the various varieties, the **Bearss** and the **Tahiti** or **Persian limes** are grown in California, while the **Mexican** or **Key lime**, a smaller variety, is grown in Florida and the West Indies.

Limes keep well in the refrigerator for up to eight weeks, but lose their perfume if kept for too long. Choose specimens that are firm and shiny-skinned.

There are various related citrus species which are not true limes and which are occasionally encountered under that name. (As with many other citrus fruits, names seem to be interchangeable, and some true limes are called lemons.) The **Indian sweet lime** (*C. limettoides*) is a rather insipid fruit, which can be used in sweet drinks without additional sugar. The **Kaffir lime** (*C. hystrix*) is not a lime at all; it belongs to a sub-species of the citrus family called *Peapeda*, and is not eaten, but is used in Sri Lanka and Indonesia as an insect repellent and in hair-rinsing water. It is grown as a houseplant or garden ornamental (weather permitting) in Europe and the U.S.A., as is the **Rangpur lime** (*C. Limonia*), an Indian variety with edible fruits. The **Spanish lime** (*Melicocca bijuga*), known in Florida, is not a lime at all, but has a similar flavour. For **Philippine lime**, see CALAMONDIN.

Limes are an important ingredient in both sweet and savoury cooking in all the regions in which they are grown, and are almost *de rigeur* in the long cool drinks of the tropics. In the West Indies and South America, raw fish and shellfish are marinated in a strong mixture of lime juice and chillis, then eaten as an appetizer. Slices of lime make an elegant decoration on sweet and savoury dishes.

# LIME CREAM

| | | |
|---|---:|---:|
| 4 tablespoons sour cream | 4 tbsp | 4 tbsp |
| 4 tablespoons whipped cream | 4 tbsp | 4 tbsp |
| ½ teaspoon sugar | ½ tsp | ½ tsp |
| juice and zest of 1 lime | | |

Combine the ingredients in the order given and beat until smooth. Use as a salad dressing or a sauce for desserts.

**Dieter's Tip**   For a low-fat substitute for the above recipe, use thick yoghurt instead of sour and whipped cream and omit the sugar. Beat in a blender.

# *Limetta* *Citrus limetta*

Sweet lemon

These fruits resemble lemons in every respect except one — you can eat them raw without puckering your face! Their mild, sweet juice tastes just like home-made lemonade, without the hard work. There are three varieties of limetta of which **millsweet**, which is grown mainly in Italy and California, is the best known. All are distinguished by having a prominent 'nipple' at the end opposite the stem, with a furrow round it.

Limettas are not normally available commercially, but occasionally fruit importers will include them in a consignment of other fruits. They make delicious lemonade, lemon curd, and lemon pie and cake filling.

This curd can be used as a pie filling (delicious in lemon meringue pie) or served as a sauce with ice cream or pancakes.

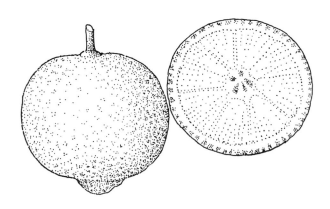

# LIMETTA CURD

| 6 limettas | | |
|---|---|---|
| 8 oz sugar | 250 g | 1 cup |
| 6 eggs | 6 | 6 |
| 4 oz butter, cut into pieces | 100 g | ½ cup |

Grate the limetta rinds and squeeze the juice. Discard any pips. Combine the juice and rinds.

Beat the eggs and sugar together until light and fluffy. Beat in the limetta rind and juice. Transfer the mixture to a bowl or the top half of a double boiler. Cook over boiling water, adding a piece of butter and stirring well until it has dissolved. Continue cooking until the mixture thickens, but do not let it boil or it will curdle.

Pour the mixture into hot, sterilized jars. Cool and store in a dark, dry place.

Makes about 2¼ pints (1.5 litres/6 cups)

# *Loganberry* *Rubus loganobaccus*

This ovoid red berry, which looks something like a large raspberry, is named after Judge J. H. Logan, the original breeder of the plant. It is a hybrid, a cross between a blackberry and a raspberry, developed in California in 1881.

Loganberries grow prolifically, and can be eaten either fresh with cream like raspberries or in puddings and pies like blackberries.

The **boysenberry** is a similar species, developed in California in the 1930s, but with a darker-coloured fruit. These berries are very delicious eaten raw in sweetened yoghurt or with cream. They are in season throughout the summer.

*Right: berries, from top to bottom: loganberries, blackcurrants, raspberries (for comparison), redcurrants and taeberries (a raspberry-loganberry cross).*

# LOGANBERRY SYRUP

| 5 lb longanberries | 2.5 kg | 5 lb |
|---|---|---|
| 1¼ pints cider vinegar | 750 ml | 3 cups |
| about 5 lb sugar | 2.5 kg | 5 lb |

Place the washed fruit in a stone crock and pour the vinegar over it. Cover the crock with two layers of muslin or cheese-cloth. Leave in a dark, cool place for three days, stirring twice daily. Strain the mixture through a jelly-bag or cheesecloth without crushing the fruit. Measure the juice into a pan, allowing 1 lb (500 g/2 cups) sugar for every 16 fl oz (500 ml/2 cups) juice.

Heat, stirring only until the sugar has dissolved. Bring to the boil, reduce the heat and simmer for 10 minutes, uncovered. Bottle in sterilized jars and seal. Refrigerate or store in a cool, dry place. Use diluted as a drink or as a syrup to pour over ice cream or pancakes.

Makes about 5¼ pints (3 litres/3 quarts)

# *Longan* *Euphoria longana*

Dragon's eye
Longyen
Lungan

At first sight, these fruits might be mistaken for LYCHEES, and in fact they are close relatives, but the skin is pale brown and not as warty, and the flesh is more translucent. In fact, the skin is orange and downy when picked and turns brown only after 24 hours. The juicy flesh is similar to that of a lychee, and it has similar brown seeds.

The longan grows in southeast Asia, and is exported mainly from India. It is in season in late summer, and should be used in the same way as the lychee.

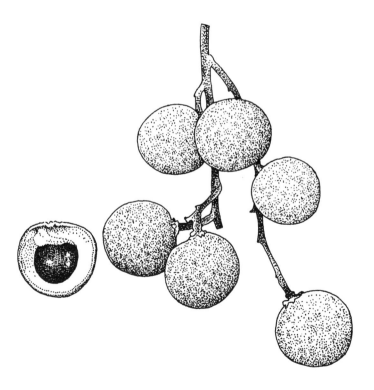

# *Loquat* Eriobotya japonica

Japanese medlar
Nispero

These round, pale-orange fruits grow in clusters on a low bush which is often cultivated for ornamental purposes rather than for fruit. Although it has long been a popular fruit in China, Japan and the Middle East, it has only recently been cultivated commercially in Spain and Israel for European and North American tables. The name 'Japanese medlar' is confusing, for sellers often call the fruit MEDLAR, though there is no relationship to that fruit at all, and any resemblance is purely superficial.

The loquat has two or three shiny brown seeds in the centre of its pale-orange flesh. The seeds are enclosed in a fibrous membrane which should be discarded before eating. The thin outer skin has a furry bloom and is sometimes freckled with tiny red spots; it peels away easily. Some varieties are too tart to eat raw, but new cultivars from Spain are sweet enough to eat without cooking. In the Middle East, the loquat is made into jams and preserves and eaten as a stewed fruit. It combines well with citrus fruit. Loquats go well with chicken and fish dishes.

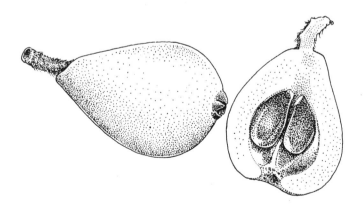

*Right: lychees at top (see page 74), with longans (centre, page 71) and rambutans (bottom, page 110): three closely related fruits.*

# *Lychee* Litchi sinensis

Leechee
Litchi (old-fashioned)

These fruits were so popular in ancient China that a poet boasted of 'limiting' himself to eating only 300 a day, while others, he claimed, ate as many as a thousand! In fact, eating too many lychees can cause stomach upsets, so it is wise to limit oneself to even fewer than 30.

Today, the lychee is more popular in southern China even than the other delicious native fruits, such as the peach, the sweet orange and the tangerine. The very first book ever written about fruit culture was devoted to lychee-growing; it appeared in 1056.

The lychee has a tough brown, scaly skin, sometimes tinged with red, which peels off easily to reveal the juicy white flesh underneath. In the centre are one or more brown seeds. The fruits are sometimes sold dried in Chinese food stores, when they are known as 'lychee nuts'; they are also sold canned. The lychees which are the standard dessert in Chinese restaurants are usually the canned variety, but the subtle flavour — a little like that of a muscat grape — can only be appreciated when the fruit is eaten fresh. The skin should be firm and dry (to show that the juice has not escaped), and should not be bruised.

Lychees are now grown in the southern U.S.A., South Africa and New Zealand, as well as in the Far East.

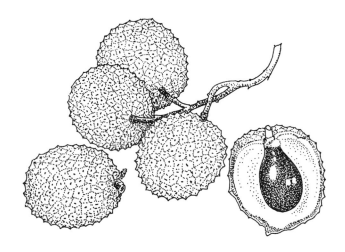

# LYCHEE ICE CREAM

| | | |
|---|---:|---:|
| *1 × 14½ oz can evaporated milk, refrigerated overnight* | *410 g* | *¾ lb* |
| *4 oz icing (confectioner's) sugar, sifted* | *125 g* | *½ cup* |
| *¼ teaspoon ground cloves* | *¼ tsp* | *¼ tsp* |
| *½ teaspoon vanilla essence (extract)* | *½ tsp* | *½ tsp* |
| *1½ lb fresh lychees, peeled, seeds removed, coarsely chopped* | *750 g* | *1½ lb* |
| *Fresh lychees for decoration* | | |
| *Slices of lime* | | |

Whisk the evaporated milk until it is thick and frothy. Then whisk in the icing (confectioner's) sugar, ground cloves and vanilla essence (extract). Fold in the chopped lychees.

Pour into an ice-cream maker and follow manufacturer's instructions for freezing. Alternatively, pour into a ring mould and freeze for 4 hours, beating once an hour to break up the crystals. Serve with extra fresh lychees and slices of lime.

6 servings

# Mamey apple _Mammea americana_

Mamee Apple
Mamey

The almost-spherical mamey apple grows on a tropical evergreen tree and is native to Central and South America and the Caribbean. The fruit has a rough golden-brown skin and orange flesh, and can be eaten raw or preserved. It can be made into jam, but is low in pectin and needs limes or hog-plums to make it set.

_Right: mangosteens, purple with leaves at top (see page 83), with two types of passion fruit (page 91): the traditional crinkly-skinned brown ones, and a rarer yellow variety._

# Mandarins, Tangerines, Tangelos, Tangors *Citrus sp.*

Mandarins (*Citrus reticulata*) are relatives of the sweet orange, though they have a distinctive flavour and a loose skin that peels away from the flesh more easily. They originated in China, and were among the oldest cultivated fruits in the Far East. The mandarin has been widely developed in Japan, where numerous varieties are found.

The enormous number of mandarin or tangerine varieties on the market (the two names are used almost interchangeably; the two varieties have the same Latin name and are indistinguishable to the layman), all sold under their varietal names, is extremely confusing. Among the most common are the **ortanique**, a large 8-inch (20 cm) Jamaican variety, and the **satsuma**, a small, mid-winter fruiting variety of Japanese origin, which is almost seedless, and is grown mainly in Spain for export to Europe. **Wilkins** or **Wilkings** is the smallest commercially grown tangerine, only about 2 inches (5 cm) across, and is also exported from California, Morocco, Brazil and Spain. The **kinnow** is a similar fruit, developed in California in 1915.

The **clementine** is the standard Mediterranean type of tangerine (though the tangerine itself originates from India). It was first discovered growing in Algeria, which is

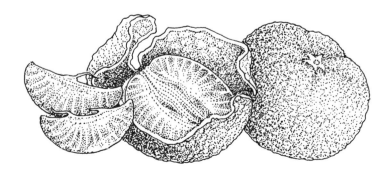

still a major exporter, but France, Israel and Morocco are now competitors. The clementine is a large tangerine, which is seedless when grown commercially. The **beauty** is a large Australian variety, with a shiny orange skin and lots of seeds. The **Dancy**, developed in Florida and named after Colonel Dancy is the most important U.S. mandarin variety.

**Tangors** are a cross between an orange and a tangerine, and bear most of the characteristics of the latter. The **temple** is a well-known type, originally bred in Jamaica and now grown mainly in Florida and Israel; it has a distinctively darker-orange skin. The **king** and, according to some authorities, the clementine, are both considered to be tangors.

**Tangelos** are a cross between a mandarin and a grapefruit, or between a mandarin and a POMELO. The **minneola** is the most distinctive variety, having a knob or boss at one end of the fruit, and bright orange skin. The minneola was developed in Florida, and is exported from Florida, California and Israel. The **orlando**, a tangelo with a skin which clings quite tightly to the fruit, is a cross between a grapefruit and an orange, and named after Orlando, Florida, where it was developed.

The **ugli fruit** is a cross between a mandarin and a grapefruit which occurred spontaneously in Jamaica and is now marketed commercially. The fruit is indeed uglier than most citrus, having a rough, warty skin which is often discoloured, being yellowish-orange in colour. The name is said to have been given to the fruit in Canada, which received the first exports. The name stuck, and was subsequently patented by the largest grower in Jamaica.

Tangerine, tangor and tangelo juices have a stronger, more distinctive taste than that of oranges, and make an excellent substitute for orange juice in any recipe. However, in some varieties the skin is too friable to make good zest, since the bitter pith tends to grate off with it, and it lacks the flavour of cooked orange rind. For this reason, tangerines do not make good marmelade.

It is hard to judge from the outside how juicy tangerines are, but they should always be orange-coloured when bought, unlike sweet oranges, which in the tropics tend to 're-green', that is, turn back to green from orange when very ripe. The **Cuban orange** is a good example of this phenomenon. All tangerines are at their best in the northern winter, even those from the West Indies.

# *Mango* *Mangifera indica*

Mangga

This deliciously juicy tropical fruit has long been prized in its native India. The Moghul Emperor Akbar planted a vast orchard of 100,000 mango trees in northern India in the late sixteenth century; since then, the mango has been cultivated intensively on the Indian sub-continent as well as in a variety of tropical and sub-tropical habitats in the Old and New Worlds.

The mango has a thin skin which is green when unripe and ripens to yellow with a pink or red blush. Some varieties are deeper pink. The flesh is very juicy with a delicate flavour. The only inconvenience in the mango is the difficulty of extracting its large, flattened seed, which adheres to the flesh by means of fibres. For this reason, the mango is rather messy to eat, and some say the best place to eat a ripe mango is in the bath!

To prepare a ripe mango for eating reasonably elegantly, it should be thinly peeled with a knife, then the flesh cut into small cubes by crisscrossing it with a sharp knife. It can then be cut away from the seed.

The bright-orange flesh of the mango is rich in carotene and vitamin C, and contains traces of potassium and other minerals. Ripe mango slices are delicious eaten with cottage cheese or ricotta.

Green mangoes can be cooked and are eaten like vegetables in India, where they are also made into a delicious chutney. The green mango is also dried and the flesh ground into a powder called *amchoor*, which is used to impart a sour flavour to food. Ripe mangoes are usually eaten raw in fruit salads or puréed to make a delicious drink.

Mangoes can cause an allergic reaction in some people, especially if the juice touches the skin; it is therefore wise to wear gloves when peeling mangoes. Mangoes are exported from India, California, Kenya and Israel.

*Left: mangoes at various stages of ripeness, showing also how the flesh can be diced.*

## CARIBBEAN GREEN MANGO DIP

| 6 small green mangoes | | |
| --- | --- | --- |
| 1 small onion | | |
| 2 tablespoons oil | 2 tbsp | 2 tbsp |
| 1 teaspoon salt | 1 tsp | 1 tsp |
| 1 teaspoon chilli powder | 1 tsp | 1 tsp |

Peel the mangoes and grind the flesh in a food processor or with a pestle and mortar, together with the onion. Beat in the oil and the rest of the ingredients. Chill before serving with fish or beef stew.

# $\mathcal{M}$angosteen  *Garcinia mangostana*

The mangosteen grows throughout southeast Asia, and is now being cultivated commercially in Central America and Australia. The tree on which it grows is extremely beautiful, but the fruits themselves are not outwardly very attractive, being of a dull purplish-brown with four large sepals around the stem. However, this thick skin peels away easily to reveal a pearly-white flesh, divided into segments like an orange, but with the flavour and texture of a LYCHEE. There are one or two pips in each section.

The mangosteen belongs to a family called the *Guttiferae*. Another member of this family, the San Domingo apricot (*Mammea america*), grows wild in Central America and is being cultivated experimentally in Florida. The San Domingo apricot has the same leathery skin as the mangosteen, but the flesh is yellow and tastes something like an apricot. It has three large pips.

The mangosteen skin is always discarded before eating; it contains a lot of tannin and is used in Asia to cure leather. The flesh is eaten raw or in fruit salads. Buy firm specimens. They are always expensive outside their native country, and are in season in summer and autumn.

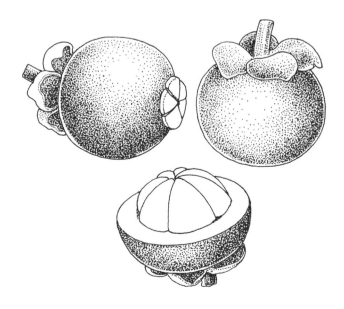

# *Medlar* Mespilus germanica

This brownish fruit, native to central and southern Europe, but also grown in northern Europe, especially the Netherlands, has two unusual characteristics. The first is that due to the strangely shaped calyx, the seeds are visible from the outside of the fruit when looked at from the top.

The medlar's other peculiarity is that it is only edible when rotted, or 'bletted', as the process is known. The fruit is picked in November, before the first frosts, and laid on straw for several weeks to decay. The flesh then becomes sweet and soft enough to eat with a spoon.

Medlar trees are now grown as ornamentals (they are relatives of the hawthorn and dog-rose), though the fruit was popular in the Middle Ages, probably because it was available at a time of year when fresh fruit was scarce. At present, medlars are not widely available commercially, but can occasionally be bought in markets. They are best eaten raw, or stewed and served with cream or custard.

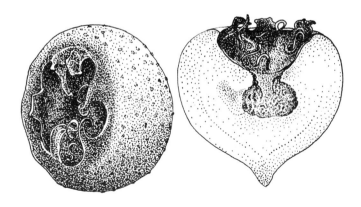

# _Melon_ _Cucumis melo sp._

Melons originated in Africa, but are now grown throughout the tropical and sub-tropical regions of the world. They are members of the same family as pumpkins, cucumbers and squashes, and like them, have juicy flesh under a thin, inedible skin, and lots of seeds in the more-or-less hollow centre of the fruit. However, some varieties which are grown for export are seedless.

Most melons are sweet, though the **Chinese winter melon**, a green-skinned variety with pale-green flesh, is used as a vegetable. Melons, like squash, fall into two kinds, summer and winter. Winter melons are thick-skinned and generally ovoid, about the shape of a rugby ball or an American football. These melons are also known as **casaba** melons (_C. melo_ var. _inodorus_). The **honeydew** melon is a member of this type. As the Latin name implies, these melons have no smell, and consequently it is very hard to determine their ripeness. One must basically choose a firm, unblemished species and trust to luck.

The summer melons are much easier to deal with. These include all the 'netted' varieties, i.e. those that look as though they are covered in brown netting, such as the **Charentais, canteloupe, Persian, ogen** and **galia** melons, whose flesh is either green or orange. When ripe, these melons have a rich, heady perfume, which can be smelt only if one holds the fruit close to the nose. They also give

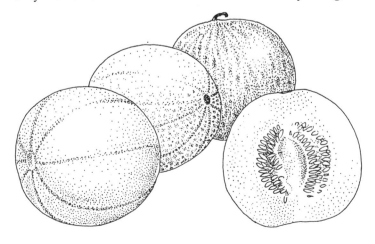

slightly when pressed at the stem end. They vary in size from small enough to be eaten by one person to a five- or six-portion melon, depending on variety and origin.

Melons must not be allowed to overripen. Always discard any whose flesh has darkened or which have become bitter, since they are potentially dangerous. They cannot be cooked but are delicious with cold meat, as well as in desserts.

The **watermelon** (*citrullus lanatus*) is a distinct variety of African origin. It is generally large and heavy, weighing between 5 lb (2.5 kg) and 10 lb (5 kg), and having a bright-red pulp containing numerous black seeds. Some varieties are oblong, with skin that can vary in colour from pale green to dark green, or alternating dark- and light-green stripes. Other varieties are completely round, and the skin of these is usually dark green, though some are striped. The flesh is mostly water. It is even possible to insert a spigot into a watermelon as one would in a barrel, and draw off the liquid.

Varieties whose flesh is bright yellow are now becoming popular. A variety whose flesh looks almost like pineapple, and is consequently known as the **pine-melon**, is now exported from Israel. The variety grown is small enough to be eaten in two portions.

The only way to tell if a watermelon is ripe is to rap it with the knuckles. It should make a muffled, as opposed to a hollow, sound. However, this is no guarantee that the flesh is really sweet. The best time to buy watermelons is from July to September; it is not a good idea to rush out and buy them as soon as they appear, as the first shipments may have been picked too early.

Hollowed-out watermelons make attractive containers for fruit salads. The seeds can be removed and the flesh chopped, mixed with white wine, vermouth, white rum or gin and chilled for a delightful cooling drink.

The **citron-melon** is a melon that is the true 'missing link' between a melon and a pumpkin. It is native to the United States, and is known as *citrouille* in France, where it is thought of as a pumpkin. In the northern United States, a thick-rinded variety is grown which has very little pulp, and is used to make a sweet, crunchy pickle. In both France and the U.S.A. it is candied and used as a substitute for the true CITRON — hence the name, citron melon. However, in the southern States the same melon is cultivated with a thinner rind and thicker, sweeter flesh, and this is eaten as a melon.

# _Nectarine_ _Prunus persica_

The nectarine is a natural mutation of the peach, and occasionally a peach tree will produce a nectarine or two spontaneously. Nectarines have a smooth skin which is yellow with a dark-red blush. The flesh is slightly firmer than that of the peach and almost always golden, though white- and red-fleshed varieties are known. Nectarines, like peaches, originated in China and may have been known to the Romans. They were first encountered in Persia by Europeans, hence the Latin name, though they were very rare in Europe until the sixteenth century.

Nectarines are rich in vitamins A and C. As with peaches and pineapples, they should be allowed to ripen on the tree, though unripe nectarines can still improve if they are left in a warm place for a few days. Nectarines should be treated like peaches for culinary purposes. They are in season in high summer. The biggest exporters of nectarines are the United States (California) and Italy.

# NECTARINE CHICKEN

| | | |
|---|---|---|
| 6 chicken joints, skinned | | |
| 1 teaspoon salt | 1 tsp | 1 tsp |
| 1 teaspoon black pepper | 1 tsp | 1 tsp |
| 1 teaspoon dried mint, | 1 tsp | 1 tsp |
| or 1 tablespoon chopped fresh mint | 1 tbsp | 1 tbsp |
| ½ pint chicken stock (broth) | 300 ml | 1¼ cups |
| 2 firm bananas or plantains, peeled and sliced lengthwise | | |
| 4 nectarines, peeled and quartered | | |
| 1 tablespoon lemon juice | 1 tbsp | 1 tbsp |
| 1 tablespoon soy sauce | 1 tbsp | 1 tbsp |
| fresh mint leaves for decoration | | |

Preheat the oven to 350 °F (180 °C/Gas Mark 4). Sprinkle the chicken joints with salt, pepper and mint and arrange them in a casserole. Pour the stock into the casserole to barely cover the chicken. Cover the pot and place it on the middle shelf of the oven. Cook for 1 hour.

Add the nectarines and bananas and sprinkle with the lemon juice and soy sauce. Cover the pot and cook for another hour. Serve garnished with mint leaves and eat with plain boiled rice or noodles.

4–6 servings

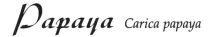

# Papaya *Carica papaya*

Pawpaw (South Africa, Great Britain)
Paw paw

This is a pear-shaped fruit (though it is much larger than a pear), and it has a thick layer of reddish-orange flesh under a thin skin, and a hollow centre filled with numerous little black seeds. The seeds are usually discarded, but have a pleasant peppery flavour and can be used in breads and as a condiment in salads. The papayas hang from large bushes, not unlike banana palms. The exact origin of the fruit is uncertain, though it first came to the attention of Europeans in the Dutch East Indies. However, the experts believe it is native to tropical America. Papayas are grown in California, Florida and Mexico and exported to Europe from Brazil and India.

A ripe papaya has a yellow skin, sometimes with patches of green. It should always be firm to the touch. The skin is not eaten.

The flavour of the fruit is extremely hard to describe, but is something like that of a melon, and is said to be an

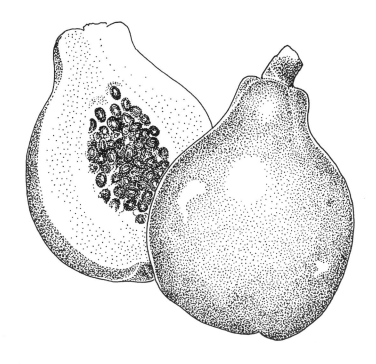

acquired taste. Ripe papaya is usually eaten raw, with a sprinkling of lime juice, and makes an especially good breakfast dish instead of grapefruit. Green papayas are baked like CHAYOTE or SQUASHES (see the companion book, *Exotic Vegetables A–Z*).

The papaya is not only highly nutritious but has various medicinal properties, which are also to be found in the leaves and bark of the papaya tree. Papaverin, extracted from the fruit, is a powerful painkiller, and other properties in the fruit are used to make contraceptive pills. The fruit also contains a powerful enzyme which tenderizes meat.

However, papayas can cause an allergic reaction in some people, especially if the juice touches the skin; it is therefore wise to wear gloves when peeling them.

## TROPICAL PAPAYA DESSERT

| | | |
|---|---|---|
| 2 ripe papayas | | |
| 12 fl oz unsweetened coconut milk | 375 ml | 1½ cups |
| 8 oz brown sugar | 250 g | 1 cup |
| 3 tablespoons lime juice | 3 tbsp | 3 tbsp |
| juice of ½ an orange | | |
| ½ teaspoon cinnamon | ½ tsp | ½ tsp |
| ½ teaspoon vanilla essence | ½ tsp | ½ tsp |
| 2 egg whites | | |

Cut the papayas in half and scoop out the seeds. Remove the flesh from the skins and purée it in a food processor. Warm the coconut milk in a pan and stir in the sugar until it has dissolved. Combine the papaya, fruit juices, cinnamon and lemon juice in a bowl. Stir the coconut milk mixture into it.

Whip the egg whites until they are stiff and fold them into the mixture. Chill before serving.

4 servings

NOTE There is absolutely no connection between the papaya and the American fruit which grows as far north as New York and Kansas and which is known locally as a **pawpaw**; this is a member of the *Anona* family (see CHERIMOYA), and should be treated in the same way.

# Passion fruit *Passiflora edulis*

Granadilla
Maracuja

A native of Brazil, the passion fruit is now produced in Australia and East Africa too. The name derives from the beautiful flower whose stigma, stamens and sepals are said to represent the instruments of torture associated with Christ's Passion. The name *granadilla* means 'little pomegranate', and the general structure of the fruit does resemble that of the POMEGRANATE.

The passion fruit has a thick purplish-brown rind enclosing a pulp full of seeds which are often eaten with the fruit. The fruit is round and about 3 inches (7 cm) in diameter. As the pulp ripens, the fruit tends to shrivel, so look deliberately for fruits which are shrunken or shrivelled, as they will be the ripest.

Passion fruit pulp is often strained and used in drinks and sauces. It is very good in fruit salads and mixed with whipped cream or ice-cream. In Australia, passion fruit is often incorporated into cakes and features in the ice-cream dessert known as a Pavlova.

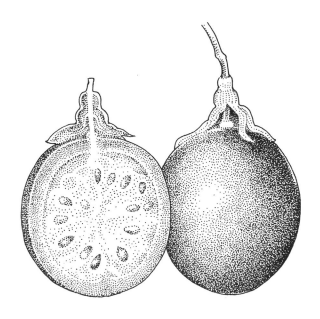

Other closely related varieties are also edible. The **sweet granadilla** (*passiflora edulis* v. *flavicarpa*), confusingly called the **giant granadilla** commercially, is larger than the passion fruit, and the skin has an attractive golden colour in contrast to the rather drab skin of the passion fruit. The seeds are flatter and more oval, and have a deliciously crunchy texture. The skin does not shrivel when the fruit is ripe, so look for fat, sleek, shiny fruit. The **banana passion fruit** (*Passiflora molissima*) is an oval variety which is coming into commerical cultivation. The flesh is said to taste even better than that of the other passion fruits. The **giant granadilla** (*Passiflora quadrangularis*), which grows in South America and Indonesia, is yellow or yellowish-green and rather acidic, but it is used to make the popular liqueur known as *maracuja*.

Many recipes require the pulp to be strained and the seeds removed, but this is a fiddly job, and canned juice can be substituted equally well because, unlike many other tropical fruits — guava for instance — passion fruit does not lose its taste when it is canned. This recipe does not require the seeds to be discarded, as they add an unusual crunchy texture to the fruit.

## PASSION FRUIT SORBET

| 3–4 passion fruits | | |
|---|---:|---:|
| 1 lb brown sugar | 500 g | 2 cups |
| 1 tablespoon liquid honey | 1 tbsp | 1 tbsp |
| 4 fl oz port wine | 125 ml | ½ cup |
| ¼ teaspoon salt | ¼ tsp | ¼ tsp |

Cut the passion fruits in half lengthwise and scoop out the pulp and seeds. Discard any pith or inner skin. Add the rest of the ingredients. Measure the mixture and add 1¾ pints (1 litre/4 cups) water to it. Stir well. Pour into freezing trays or an ice-cream maker. If using freezing trays, beat the mixture at least once (preferably twice) as it begins to freeze, to break up the crystals.

8 servings

# $\mathcal{P}$ersimmon  *Diospyros kaki*

Date plum
Kaki fruit
Sharon fruit

This beautiful, brilliant-orange fruit, which looks a little like a tomato with a large, dried green calyx, is unusually rich in vitamin A, and quite a good source of vitamin C. Specimens vary in shape, depending on where they are grown, from completely round to pointed at one end. The fruit grows on the tree at a time when it is bare of leaves, in late autumn and winter, making it look as though the tree is hung with lanterns.

The persimmon originates from China or Japan and both countries grow a large number of different types of the fruits. There is also a closely related American variety, whose Latin name is *Virginiana*, which is also known as the **Virginian date**. Nowadays, persimmons are grown in all fairly dry, sub-tropical climates, including California and the Middle East.

The flesh of the persimmon is almost like a firm jelly. The skin is very thin and can be eaten or not, as desired. Many varieties have several brown seeds, which are discarded before eating, but the Israeli cultivar known as the **Sharon fruit** is seedless.

Persimmons are absolutely delicious when completely ripe, but for a long time even the cultivated varieties tasted like a bitter mouthful of fur when not absolutely mature. This was due to the tannic acid in the unripe fruit, and fruit which had not ripened evenly would taste unpleasant in parts. The Sharon fruit and other new

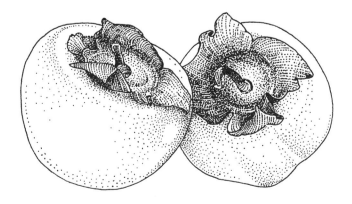

varieties have overcome the problem, though it is always advisable to wait until the fruit is completely ripe before eating it.

Persimmons can be ripened by keeping them in a warm place in a brown-paper bag. Choose fruit which is un-damaged and as ripe as possible, which means it should be tender to the touch and slightly translucent. Very firm flesh means the fruit is not ripe.

Persimmons are eaten raw, and incorporated into a large variety of sweet dishes such as cakes, puddings, ice creams and fools. They are too sweet to go well with meat or fish. If ripe persimmons are placed straight in the freezer the flesh freezes very well, and any remaining traces of tannic acid are removed by freezing. They then merely need to be thawed out a little in order to be perfect for eating as a frozen dessert.

The **black sapote** (*Diospyros ebenaster*), which is no relation to the WHITE SAPOTE, is closely related to the persimmon. It is cultivated mainly in Mexico and the West Indies, and bears a slight resemblance to the per-simmon in shape, though the fruit is slightly ribbed. The skin is olive-green and the flesh chocolate-brown. Its coloration may be the reason why it has not become popular outside its native environment even though the flesh is as sweet as that of the persimmon.

### PERSIMMON FROZEN YOGHURT

| | | |
|---|---|---|
| 2 *ripe persimmons* | | |
| *8 oz sugar* | *250 g* | *1 cup* |
| *8 fl oz water* | *250 ml* | *1 cup* |
| *1 tablespoon liquid glucose or white corn syrup* | *1 tbsp* | *1 tbsp* |
| *16 fl oz plain yoghurt* | *500 ml* | *2 cups* |
| *pinch of cinnamon* | | |

Remove the calyxes from the persimmons, and any seeds inside the fruit. Purée it in a blender or food processor. Put the sugar and water in a heavy-based pan. Bring to the boil without stirring and when the sugar has completely dissolved add the glucose or white corn syrup. Boil briskly for 5 minutes.

Combine the fruit purée with the yoghurt and add the cinnamon. Pour the syrup into the mixture, beating hard. Freeze in freezer trays in the refrigerator or in an ice-cream maker. If freezing in the refrigerator, stir the mixture once every 30 minutes for 3 hours, to break up the crystals.

4–6 servings

# *Physalis*  *Physalis peruviana*

Cape gooseberry
Chinese lantern
Goldenberry
Ground cherry
Peruvian cherry

The round berry of this fruit is encased in a papery calyx, giving it its name of Chinese lantern, which is more often applied to the ornamental garden plant. The name Cape gooseberry used to be most commonly applied to it, but is falling out of favour for political reasons, though it is still widely grown and exported from South Africa. The berries are bitter when green, but ripen to bright orange, when they have a spicy, sweet flavour. The skin is smooth and shiny, and contains numerous edible seeds.

As is clear from its Latin name, the physalis comes from Peru. It is unusually rich in vitamin A, and is quite a good source of vitamin C. It is in season in late autumn. Choose unblemished fruits which still have their calyx.

Physalis can be eaten raw, but are very often included in jams, preserves and fruit salads. They make particularly attractive dipped and candied fruits, and can be eaten with other dipped fruits or used as cake decorations. Try and keep the papery calyx in place, since this enhances the appearance further.

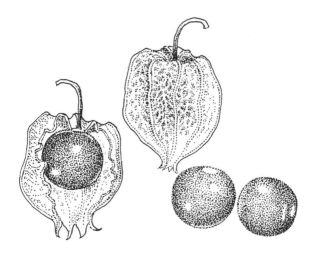

# FONDANT-DIPPED PHYSALIS

| | | |
|---|---|---|
| 8 fl oz sugar | 250 ml | 1 cup |
| 5 tablespoons water | 5 tbsp | 5 tbsp |
| 2 tablespoons liquid glucose or light corn syrup | 2 tbsp | 2 tbsp |
| 1 lb physalis | 500 g | 1 lb |
| extra sugar for dipping | | |

Put the sugar, water and glucose or corn syrup into a heavy-based pan. Bring to the boil without stirring and boil until the mixture registers 236 °F (113 °C) on a sugar (candy) thermo-meter, or until a drop of it forms a soft ball in a glass of cold water.

Wet a cold work surface (preferably a marble slab). Pour the syrup on to it, and with a pastry scraper or strong palette knife quickly begin scraping and turning it until it becomes cloudy. Continue to work it until it is completely opaque and has a crumbly texture, about 7 minutes. It should then be cool enough to handle.

Wet your hands and form it into a smooth ball. It can then be stored indefinitely in the refrigerator until required.

Line a tray with wax paper. Wash the physalis fruits carefully, without removing the stem or calyx, and pat them dry. They must be absolutely dry on the outside or the fondant will not stick to them.

To melt the fondant for coating, put it into a bowl and place it over gently simmering water. Stir it constantly until it has the consistency of very thick cream.

Put the sugar for dipping into a bowl. Dip each fruit in the fondant, holding it by the stem, and do not immerse it com-pletely in the fondant, but leave a little of the orange colour of the fruit showing. Hold it over the fondant bowl until it stops dripping. Then dip the fruit in the sugar, twirling off the excess. Leave it to dry and harden on the wax-paper tray.

Store the fruits in an airtight tin lined with wax paper. They should be refrigerated and will not keep longer than three days.

Makes about 30

NOTE Fondant powder, which simply needs mixing with water, can be bought ready-made from specialist stores supply-ing home candy-making equipment.

# *Pineapple* *Ananas comosus*

The pineapple, with its magnificent crown of leaves, has always been considered the king of fruits, and has graced the tables of the rich and famous in the western world ever since its introduction into Europe from South America in the seventeenth century. The name derives from its superficial resemblance to a pine cone, but there is absolutely no relationship. The pineapple is a member of a family of tropical plants called the *bromeliads*, of which it is the only edible species.

Today, pineapples are cultivated throughout the tropics, especially where the rainfall is high. Principal exporters of pineapples are Hawaii, the Ivory Coast and Dominica.

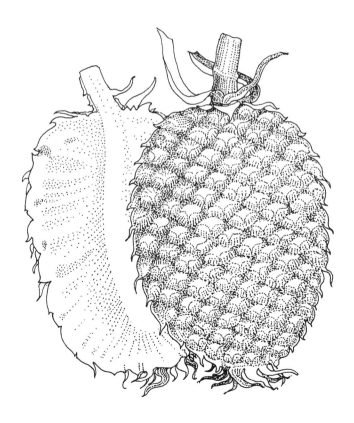

When raw, the pineapple contains an enzyme called *bromelin*, which tenderizes and decomposes protein. This is one reason why pineapple goes so well with meat and fish dishes, as well as desserts; however, it means that pineapple will 'eat' gelatine, and so cannot be made into a jelly which involves a jelling agent derived from animal sources (though it will not affect a vegetable jelling agent), unless it is first cooked. Cooking destroys the enzyme, so canned pineapple contains no bromelin. Some people prefer canned pineapple, as it is not as acid as the fresh variety when eaten outside its native country.

Pineapples were once cultivated in hothouses in Europe, though one suspects the fruit was grown in cooler climates more for its beauty than its flavour, as the fruit can hardly have been very sweet. They are unusual fruits in that they do not ripen off the bush on which they grow. A day before they mature, a rush of sweet sap shoots from the roots into the fruit, ripening it all at once, and it is only at this moment that the fruit is suitable for picking. For this reason, it is no good buying an unripe pineapple and hoping it will ripen by keeping it in a warm place for several days. The best way to tell if a pineapple is ripe before eating is by the smell. If held close to the nose, it should give off a sweet smell. The outer colour of the fruit is no indication, because some pineapples remain green even when ripe. The crown of leaves should be glossy and healthy-looking, and not withered or dry, and a leaf plucked from the crown should come away easily. The fruit should be eaten as soon as possible after it is bought. Pineapples can be bought all year round in Europe and North America and are at their best during the summer of the country in which they are grown.

To prepare a pineapple for eating, the rough skin should be sheared away with a sharp knife all over the fruit. The crown can be left on or not as desired. This will reveal a row of 'eyes' like those in potatoes, but in pineapples it will be observed that they follow a spiral pattern around the fruit. You can either remove them singly with a sharp knife or cut a spiral groove right around the fruit, removing them as you go. The hard core should also be cut away. Pineapple corers can be bought from cookware shops, or one can use an apple corer or a sharp knife.

An elegant way to present pineapple for eating is to peel the fruit without cutting off the crown, remove the eyes and lay the fruit sideways; cut it crosswise into even slices, then neatly cut the core from each slice. Fit the slices back together again to resemble the intact fruit, leaving the crown in place. The hole left by the core can be filled with fruit salad, drenched in orange-flavoured liqueur.

Another method is to make a pineapple outrigger, Hawaiian style. The fruit is peeled, leaving the crown in place, then the eyes are removed. It is then cut vertically into quarters. The core is removed from the quarters and most of the flesh is cut away from the boat-shaped slices, then cut into chunks and replaced on the slices, so that it can be eaten with a spoon.

# *Pitanga* *Eugenia uniflora*

Brazilian cherry
Cayenne cherry
Petanga
Surinam cherry

This attractive, bright-red fruit bears a strong resemblance to a cherry, and grows in clusters like cherries. It has
deep, regular furrows, however, and may contain more
than one seed.

The name pitanga is Portuguese, as the shrub on which
the pitanga fruit grows was discovered in Central America by the Portuguese, who spread it around the world.
Hitherto, this attractive plant was grown outside its
native country more for its beauty than the nutritiousness
of the fruit. The leaves are a rich, dark red and the flowers
are white. The fruits are made into jams and drinks, and
have a pleasant, slightly sour flavour. They can also be
eaten raw.

The pitanga is one of those exotic fruits that were once unknown to anyone but keen botanists and gardeners, but it is coming into commercial cultivation for export from sub-tropical countries, and it will soon be seen in the fruit markets of northern Europe and America. It is being grown in Australia, New Zealand and Israel. The pitanga, like many such new exotica, is in season in mid- to late winter, when there are few home-grown fruits on the market in colder climates.

Pitangas combine well with other fruits, such as citrus, and make an attractive garnish to cakes (use them instead of cherries for Black Forest gâteau) and to ice cream.

## PITANGA SORBET

| | | |
|---|---:|---:|
| 1 lb sugar | 500 g | 2 cups |
| 8 fl oz water | 250 ml | 1 cup |
| 1 tablespoon liquid glucose or 2 tablespoons light corn syrup | ½ tbsp | ½ tbsp |
| 1 lb pitangas, pitted | 500 g | 1 lb |
| 1 lime, juice squeezed, rind grated | | |
| 2 egg whites | | |

In a heavy pan (not one made of aluminium), such as a preserving pan, heat the sugar and water, stirring only until the sugar has dissolved. Bring to the boil and add the liquid glucose or corn syrup. Boil briskly for 5 minutes.

Remove the pan from the heat and stir in the pitangas, and the rind and juice of the lime. Cool to room temperature. Whip the egg whites into soft peaks and fold them into the mixture. Pour into freezing trays or an ice-cream maker. If using freezing trays, beat the mixture at least once (preferably twice) as it begins to freeze, to break up the crystals.

4 servings

NOTE  Liquid glucose and corn syrup are both used here to stop the mixture from crystallizing. Liquid glucose is available from pharmacists, but is hard to get in the U.S.A.; corn syrup works just as well.

# *Pomegranate* Punica granatum

The beautiful pomegranate is a fruit famous in the legends and literature of many countries. The thick skin of the fruit, which varies in colour from pale pinkish-yellow to bright red, holds a large number of seeds, each surrounded by a sac of juice called an aril. The arils are embedded in a bitter pith that should always be discarded before eating.

Pomegranates, with their many seeds, are a symbol of fertility in many cultures, including Chinese and Indian. They are the oldest symbol of Judaism, predating the Star of David by many centuries, and pomegranate-shaped silver pommels are used to cover the wooden handles of the Scrolls of the Law, which are read in the synagogue. The wild variety is said to hold 613 seeds, as many as the commandments in the Old Testament.

The pomegranate is said to have originated in Persia and spread westwards to the Middle East and eastwards to China. The juice produces an indelible dye, which has been used in the making of Persian carpets. For this reason, it is wise to ensure that the juice is kept away from valuable table linen.

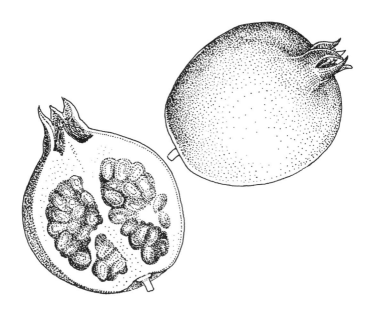

Choose pomegranates which are as brightly coloured as possible. The skin should be shiny and not withered. They are in season from late summer to early winter. The fruits keep well because of their thick skin, and they are rich in vitamin C.

To prepare pomegranates, you can cut them in half and squeeze out the juice on a lemon-squeezer, or slit through the skin lengthwise in segments, like an orange, and peel it back. Then carefully remove the pith and separate the arils. Discard the skin and pith.

Pomegranate juice makes a refreshing drink. The arils can be dried and are used extensively, in both fresh and dried form, in the cooking of India and Iran.

## CHICKEN BREAST WITH POMEGRANATE SAUCE

| | | |
|---|---|---|
| *2 chicken breasts* | | |
| *1 oz butter* | *25 g* | *2 tbsp* |
| *8 fl oz chicken stock (broth)* | *250 ml* | *1 cup* |
| *1 fresh pomegranate or 2 oz dried arils* | *50 g* | *2 tbsp* |
| *½ teaspoon lemon juice or mango powder* | *½ tsp* | *½ tsp* |
| *2 oz ground walnuts* | *50 g* | *¼ cup* |
| *pinch of sugar* | | |

Bone the chicken breasts and cut them in half. Melt the butter in a frying-pan (skillet) with a lid and cook the chicken until it is lightly browned all over.

Add the stock and the rest of the ingredients.

Cover and simmer until the chicken is tender, about 30 minutes. Serve with boiled rice.

4–6 servings

# $\mathcal{P}$omelit _Citrus sp._

The pomelit is a hybrid, a cross between the POMELO and the grapefruit, developed by the Israelis as a new export fruit. The ripe fruit has a green to greenish-yellow skin and has the slightly cone-shaped top of a pomelo, but is the size of a large grapefruit. The fibres and skin over the segments are tough, like those of a pomelo, but the flesh is sweeter than that of a grapefruit and juicier than the ordinary pomelo.

Do not be deterred by the green colour of the skin: a green skin on citrus fruit does not necessarily mean that the fruit is unripe. There is a process called 're-greening', whereby a citrus fruit which has turned yellow or orange turns green again after ripening if it is exposed to a lot of sun. For this reason many of the citrus fruits which grow in the tropics — pomelos, limes, the so-called 'Cuban' oranges, and even some grapefruits — have green skins.

Pomelits are best eaten raw for dessert. The peeled segments make an unusual raw garnish for meat and fish.

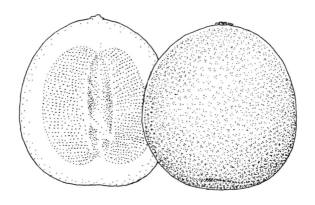

# Pomelo *Citrus grandis*

Adam's apple
Pummelo
Shaddock

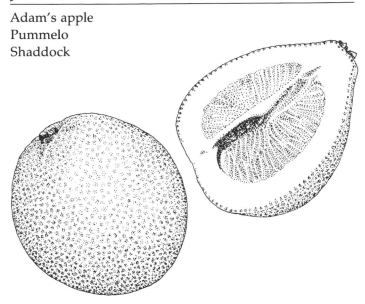

This is the largest of all citrus fruits, sometimes being as large as a man's head. It has a distinctive shape, rounded and coming to a slightly conical shape at the stem end. The skin is green or greenish-yellow even when ripe, and the flesh is divided into very coarse juice sacs. The flesh is usually green, but is pink in some varieties. The pith is thick, but separates easily from the flesh.

The pomelo originated in southeast Asia, and was until recently rarely seen outside that area. In the last few years the Israelis have decided to market it commercially, and it can now be bought all over Europe. In the United States, the Vietnamese immigrants have popularized the fruit (usually known in the U.S.A. as pummelo), because it has a symbolic significance in the Vietnamese New Year celebrations.

The pomelo is believed to be one of the parents of the grapefruit, since the grapefruit is a spontaneous hybrid which first occurred in the 1880s in Jamaica (the other parent may have been the lemon). It is not as bitter as a graprefruit and can be eaten for dessert, but is best eaten raw. Since the pith is so plentiful, the fruit is best prepared for eating by removing all skin and fibrous parts, including the thick skin covering the segments.

Pomelos should be yellow in parts when ripe. They should always be firm to the touch. They are in season in late winter.

# Prickly pear *Opuntia ficus-indica*

Barbary fig
Cactus pear
Indian fig
Sabra
Tuna (western American and Mexican)

This is the fruit of the prickly pear, which despite its name is, like all true cacti, native to the southwestern United States and Mexico. Due to its usefulness as a natural hedgerow plant, the prickly pear cactus is now grown in many arid regions, particularly the Mediterranean. The Israelis have used their name for this fruit, the sabra, as a nickname for native-born Israelis, because they claim that, like the fruit, the native-born Israelis are prickly on the outside and sweet inside.

Although non-prickly varieties of the cactus have been developed, most are still covered in long spines, which are surrounded by tiny, sharp hairs that are extremely

irritating when they touch the skin. Consequently, if you buy prickly pears with their skins on, you should wear thick gloves or use tongs to handle them, and if picking them from the cactus, do not touch the fruit at all. The best way to pick the fruit is to use an empty tin can on the end of a stick, hooking the can over the fruit. Then roll it on the ground with a stick to rub off as many spines and hairs as you can. The skin is thick and will withstand a little pressure. To get at the sweet centre, slice off either end, then make a lengthwise slit down one side with a sharp knife. Peel back the thick skin to expose the flesh.

A prickly pear is ripe when its green colour disappears and it becomes pinkish-yellow. The flesh is very sweet and interspersed with seeds, which can be sieved out or eaten with the fruit. It can be eaten raw and is delicious in a fruit salad, sprinkled with lime juice. In Arizona and Mexico it is made into candy.

All varieties of cactus fruits are edible, provided the same precautions are taken when picking them.

# *Quince*  *Cydonia vulgaris*

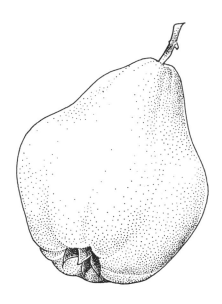

This plant is native to western Asia (Iraq and Iran), where it still grows wild. The fruit roughly resembles a large yellow pear, or sometimes an apple, and the skin is covered with furry down. The seeds are grouped in the centre as in an apple core.

The quince has a very hard, grainy flesh, which is too tough to eat raw. However, it is highly prized for its delicate flavour when cooked, and wherever it is known it is made into a variety of delicious preserves, since it is rich in pectin and jells easily. The quince turns an attractive deep pink when cooked. The Spanish and Portugese name for the quince, *marmela*, is the origin of the English word 'marmalade', since this type of preserve was originally made from quinces.

Quinces are in season in mid-autumn. They can be bought unripe and ripened in a sunny place. Always rub off the down before cutting or peeling. Quinces are much easier to cut if parboiled for 10 minutes or microwaved whole for 4 minutes. They can be stewed with sugar, or cooked and puréed, then added to a thick sugar syrup to make quince paste. In Asia, the quince is often combined with chicken or red meat and makes a delicious contrast in flavour.

# CHICKEN-STUFFED QUINCES

| | | |
|---|---:|---:|
| 4 large quinces | | |
| 1 lemon | | |
| 1 chicken breast, boned | | |
| 3 tablespoons ground almonds | 3 tbsp | 3 tbsp |
| ¼ teaspoon ground cloves | ¼ tsp | ¼ tsp |
| ¼ teaspoon cinnamon | ¼ tsp | ¼ tsp |
| 1 teaspoon rosewater | 1 tsp | 1 tsp |
| ½ teaspoon salt | ½ tsp | ½ tsp |
| 1 oz butter, cut into pieces | 25 g | 2 tbsp |
| About 8 fl oz chicken stock (broth) | 250 ml | 1 cup |

Parboil the quinces in water to cover for 10 minutes or microwave them whole for 4 minutes. Leave them to cool, then peel them carefully and slice them in half lengthwise. Scoop out the core and most of the flesh. Squeeze lemon juice over them to prevent discoloration.

Preheat the oven to 400 °F (200 °C/Gas Mark 6). Remove any fat from the chicken breast, pull or cut it into pieces and put it in a food processor. Add the ground almonds, the spices, rosewater and salt and grind to a smooth paste. Fill the quince shells with the paste, so that it covers the shells. Dot the shells with pieces of the butter. Place them in a shallow oven dish and add enough stock (broth) to come one-third of the way up the sides of the quinces. Bake for 1 hour, or until the quinces are soft and the meat is cooked through, adding more stock (broth) if necessary. Serve as a starter, or as a main course with rice.

4 servings

# $\mathcal{R}ambutan$ _Nephelium lappaceum_

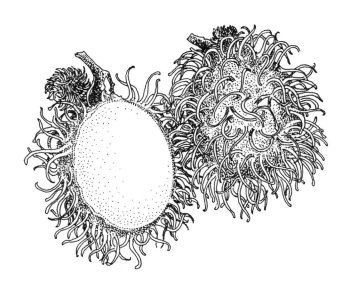

This most striking fruit, which grows in clusters, is now exported worldwide from its native southeast Asia, as well as from Australia. The name derives from the Malayan word for 'hairy', and the reddish skin looks as though it has green and red hairs growing out of it, which are in fact soft spines. Despite its fearsome appearance, the skin is not prickly and peels off easily, revealing pearly-white flesh underneath, with a shiny brown seed in the centre.

Like the better-known LYCHEE, the rambutan is often sold peeled and canned. It is usually eaten raw or lightly cooked in syrup, but its striking appearance makes it even more valuable as a decoration for the fruit bowl or among a dried flower arrangement. When buying fresh rambutans, make sure the colour is a ripe ruby red, for the skin darkens with age. The fruit keeps well when refrigerated and will last about a week, unpeeled, at room temperature. Rambutans are in season in autumn.

Here is an unusual lychee or rambutan dish, in that it is a main course. It has become popular in Indian restaurants, though it is not really authentic, since Kashmiris do not eat chicken!

# KASHMIRI CHICKEN

| | | |
|---|---:|---:|
| 4 lb chicken, cut into serving pieces | 2 kg | 4 lb |
| 1 tablespoon curry powder | 1 tbsp | 1 tbsp |
| 1 tablespoon garam masala | 1 tbsp | 1 tbsp |
| 1 teaspoon salt | 1 tsp | 1 tsp |
| 4 tablespoons fresh lime juice | 4 tbsp | 4 tbsp |
| 4 tablespoons natural yoghurt | 4 tbsp | 4 tbsp |
| ½ teaspoon ground turmeric | ½ tsp | ½ tsp |
| 1 inch ginger root, peeled and grated | 2.5 cm | 1 in |
| 2 garlic cloves, minced | | |
| 1 unripe mango, peeled and sliced | | |
| 6 rambutans or lychees, peeled | | |
| 2 oz pistachio nuts, shelled | 50 g | ¼ cup |
| 2 oz seedless raisins | 50 g | ¼ cup |
| 16 fl oz water or chicken stock (broth) | 500 ml | 2 cups |

Heat the oven to 400 °F (200 °C/Gas Mark 6). Skin the chicken pieces. Combine the curry powder and garam masala with the salt and turmeric. Mix with the lime juice, yoghurt and ginger root. Place the pieces of chicken on a roasting tray and smear them with the mixture. Cook for 1 hour.

Remove the chicken from the oven and reduce the heat to 350 °F (180 °C/Gas Mark 4). Bone the chicken. Discard the bones and cut the chicken into bite-sized pieces. Place it in a casserole. Add the rest of the ingredients and the water or chicken stock (broth). Cook for 45 minutes. Serve with plain boiled rice.

6–8 servings

# *Sapodilla* Achras sapota

Chicle
Chico
Mammee sapota
Marmalade plum
Naseberry

This oval brown fruit, with its rough skin, bears some resemblance to a potato, but it grows on the tree from which chewing gum is made. The sapodilla is native to South America, but is grown in the West Indies and the Philippines. It can only survive in the tropics, and is considered one of the finest tropical fruits, growing abundantly in these regions.

The fruit has a central arrangement of shiny black seeds in a star shape. The flesh is reddish to dark brown in colour, with a firm, finely granular texture. The flesh is extremely sweet, almost cloying — something like a PERSIMMON — and it can be treated in the same way.

Varieties related to the sapodilla are the **sapote** (*Calocarpum mamosum*), which has a large central seed similar to that of an avocado; the **green sapote** (*Calocarpum viride*); and the STAR-APPLE. The WHITE SAPOTE is not related.

The following recipe can be used equally successfully with all these fruits:

# SAPODILLA PANCAKES

| | | |
|---|---|---|
| 4 sapodillas | | |
| 5 tablespoons (all-purpose) flour | 5 tbsp | 5 tbsp |
| 2 eggs | | |
| 1 tablespoon sugar | 1 tbsp | 1 tbsp |
| 10 fl oz milk | 300 ml | 1¼ cups |
| 6 tablespoons safflower oil | 6 tbsp | 6 tbsp |

Sift the flour into a bowl and make a well in the centre. Break the eggs into the well and add the sugar. Gradually add the milk, beating constantly. Beat the batter until it is smooth. Cover it lightly with a cloth and leave it to rest for at least 15 minutes, while you prepare the sapodillas.

Peel the fruit and discard the seeds. Mash or purée the pulp, then stir it into the batter. Refrigerate the batter for 30 minutes.

Heat an omelette (omelet) pan or non-stick frying-pan (skillet). Put 1 tablespoon of the oil into the pan. When it is very hot, add a tablespoon of the batter. Fry lightly about for 2 minutes on each side. Repeat until all the oil and batter are used up. Do not add batter to the pan until the oil is hot. Serve hot, with jam or syrup, if liked.

Makes about 12 pancakes

# _Seville orange_ Citrus aurantium

Bigarade
Bitter orange
Sour orange

This is the earliest orange known to the western world, having been introduced into Europe by the Moors in the fifth century AD, and possibly even before that. The skin and flesh look like that of a sweet orange, though there are more seeds, but the juice is sour. The Seville orange is most useful in cooking as a sour flavouring or for making marmalade from the thick, bitter rind. The seeds are rich in pectin, so they jell well. Home-made marmalade is so delicious that it is well worth trying.

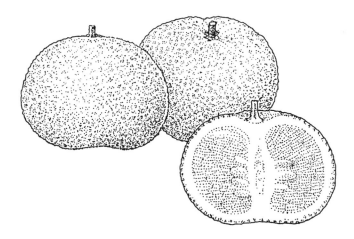

# SEVILLE ORANGE MARMALADE

| | | |
|---|---:|---:|
| 2½ lb Seville oranges | 1.25 kg | 2½ lb |
| 2 lemons | | |
| 4 lb preserving sugar | 2 kg | 8 cups |
| 4 fl oz Scotch whisky | 125 ml | ½ cup |

Cut all the fruit into quarters. Remove the seeds and place them in a piece of cheesecloth or muslin, tied tightly at the top. Scrape the pulp from the fruit and reserve it. Put the peel in a large bowl or pan; do not use aluminium. Cover with water and add the muslin bag of seeds. Leave to soak for 24 hours.

Drain the peel and seeds. Shred the peel coarsely or finely, depending on how you like your marmalade. Put the peel and cheesecloth into a preserving pan and add 3½ pints (2 litres/2 quarts) water. Bring to the boil and simmer, uncovered, for 1 hour. Remove and discard the cheesecloth.

Pour the sugar into a cake tin and leave it in the oven for 15 minutes on lowest heat. Grind the fruit pulp in a food processor and add it with the sugar to the pan of shredded peel. Stir over low heat until the sugar has dissolved, then increase the heat slightly and bring the mixture to a slow boil. Cook about 45 minutes without stirring, then start testing it by dropping half a teaspoonful on to a cold saucer. If it jells when cool, it is ready.

Remove the pan from the heat and quickly stir in the whisky. Bottle in sterilized preserving jars while still hot, filling the jars to the brim and covering the mixture with a circle of waxed paper. Seal tightly. The marmalade should be kept at least one month before eating, and it improves with age.

Makes about 4 pints (1.75 litres)

# Soursop *Anona muricata*

Bullock's heart
Guanabana

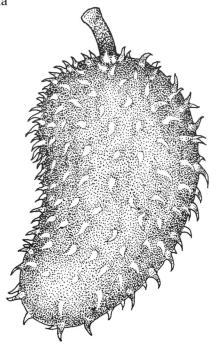

This is another member of the CHERIMOYA (custard apple) family. The soursop is about twice as big as the cherimoya and has a heart shape which gives it its alternative name. The skin is pale green when unripe, with numerous short, fleshy spines on the surface. It turns brown when ripe, and has a rank, bitter flavour so it is always discarded. The flesh has a sharper flavour than the custard apple but is still very sweet, and there are numerous black seeds.

When ripe the soursop should be brown all over and give when pressed with the fingertip. It is in season in late autumn.

The soursop makes a wonderful ice cream, and refreshing drinks too; the Cubans make a wonderful long drink with soursop called *champola de guanàbana*.

# SOURSOP SORBET

| 1 soursop | | |
|---|---|---|
| 1 lb sugar | 500 g | 2 cups |
| 8 fl oz water | 250 ml | 1 cup |
| ¼ teaspoon ground cloves | ¼ tsp | ¼ tsp |
| ¼ teaspoon ground allspice | ¼ tsp | ¼ tsp |

Peel the soursop and remove all the seeds. Purée the flesh in a blender. Put the sugar and water into a heavy-based pan and bring to the boil without stirring. Boil briskly for 5 minutes. Remove the pan from the heat and add the spices. Stir the syrup into the puréed soursop. Pour into an ice-cream maker and follow the manufacturer's instructions.

Alternatively, pour into freezer trays, place in the freezer and stir every half-hour until the mixture is frozen.

**Serving suggestion**   Serve with orange-flavoured liqueur.

# _Star-apple_ Chrysophyllum cainito

Cainito

This magnificent fruit originates from South America and the Caribbean, and is especially abundant in Jamaica and Haiti. It seems a likely candidate for export and popularizing in Europe and North America, thanks to its delicious flavour and spectacular appearance. The thin purple skin hides a thick layer of purple flesh, which changes to white in the centre. When the fruit is cut open, the glistening, gelatinous flesh shows the star-shaped pattern of the transparent seeds. The mild, sweet flesh is best eaten raw or in a fruit salad. Charles Kingsley, the novelist, likened it to an 'evergreen peach'.

The star-apple is no relation to the star-fruit or CARAMBOLA, but is related to the SAPODILLA, and can be used in the same way.

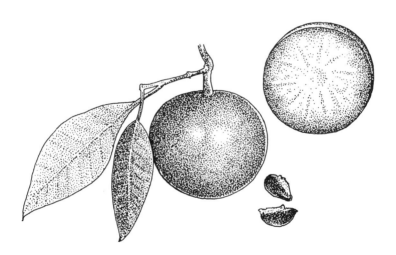

# *Tamarillo*  *Cyphomandra betacea*

Jambolan
Java plum
Tree tomato

This fruit, which has a passing resemblance to a tomato but is much more spectacular, originates from Peru. The large (5 inches or so) fruits are more oval and have a thin, tough, purplish-red skin. The inside flesh is shaded from orange to red to yellow. The small black seeds are ranged in rows like in a tomato.

When tamarillos are ripe their flesh should be fairly soft to the touch, like that of an avocado. They will keep in the refrigerator for about a week. They are delicious in salads, as a condiment with grilled steak or ham, and baked in the oven round the roast.

The tamarillo can be eaten raw, and to do so, the fruit is cut open and the flesh scooped out. The skin is fairly tough and should not be eaten. The strong acidity of the flesh means that tamarillos are usually better cooked. To cook them, immerse them in boiling water for about 3 minutes, then peel off the skin.

The tamarillo is now an important New Zealand crop, though the fruit is also grown in the U.S.A., India and Sri Lanka. It is exported to northern Europe and the U.S.A. from New Zealand.

Tamarillos are high in vitamin C (about 31 mg per 100 g) and low in calories, averaging 26 per 100 g.

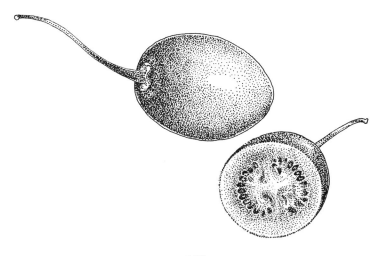

# TAMARILLO PIE

| | | |
|---|---:|---:|
| 8 oz butter | 250 g | 1 cup |
| 8 oz sugar | 250 g | 1 cup |
| 2 eggs | | |
| 12 oz all-purpose flour | 375 g | 1½ cups |
| 2 teaspoons baking powder | 2 tsp | 2 tsp |
| 4 oz bran flakes, crushed | 125 g | ½ cup |
| 6 ripe tamarillos | | |
| 6 oz chopped walnuts | 175 g | ¾ cup |
| 3 tablespoons brown sugar | 3 tbsp | 3 tbsp |
| 1 teaspoon cinnamon | 1 tsp | 1 tsp |
| 1 egg yolk, mixed with 2 teaspoons milk and 1 teaspoon sugar | | |

Grease a 9-inch (22.5 cm) pie dish. Cream the butter and sugar together. Add the beaten eggs. Sift the flour and baking powder and beat it into the butter mixture. Mix to a firm dough. Roll out half the dough and use it to line the pan. Sprinkle with half the bran flakes.

Immerse the tamarillos in boiling water for 3 minutes. Remove and skin them. Slice them carefully into rings and arrange them over the bran flakes. Sprinkle them with the rest of the bran flakes, the brown sugar, cinnamon and the rest of the cornflakes.

Preheat the oven to 350 °F (180 °C/Gas Mark 4). Roll out the rest of the dough and use it to cover the pie. Brush it with egg yolk, milk and sugar glaze. Bake for 45 minutes or until well browned. Serve hot with whipped cream or vanilla ice cream.

6–8 servings

# *Tamarind* *Tamarindus indica*

Indian date
Tamar hindi (Arabic and Hebrew)
Tamarin (French)
Tamarindo (Spanish)

The beautiful tree on which the tamarind pod grows is a native of India, but is now cultivated in most of the tropical and sub-tropical areas of the world. The brown pods are sometimes sold in semi-dried form, but more often than not as a compacted, sticky mass. When buying the compacted variety, check to see there are no telltale holes of insect penetration. A syrup is made from them which is used to flavour drinks throughout western Asia, including the Middle East, and in tropical South America.

To prepare tamarind from pods or pulp, soak in boiling water to cover, then pound lightly into a purée. Sieve the purée, to remove seeds and fibres. Use it in Indian cooking, or mix it with water to make a refreshing sweet-and-sour drink.

Tamarind pods are available in the summer and autumn, but the pulp is available year-round. They contain some vitamin C, iron and trace elements, but are high in calories, due to sugar content.

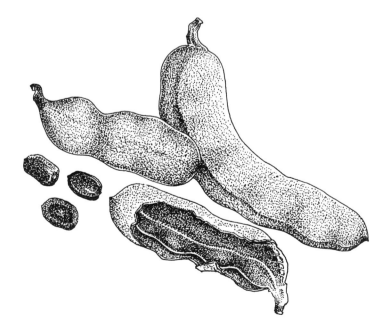

# *White sapote* Casimiroa edulis

Custard apple

The white sapote bears a strong outward resemblance to a green apple, hence its alternative name. It is not to be confused with the CHERIMOYA, however, although this is also called a 'custard apple'. Inside, the flesh is creamy and smooth, and there are four or more large white seeds in the centre. The fruit is subtropical, and grows in Mexico and California, on high ground, and can also be found in Florida.

The fruit is usually eaten fresh, though it can be made into desserts and ice creams like the cherimoya.

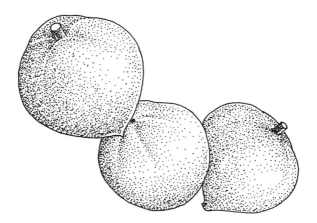